Action Against Climate Change

The Kyoto Protocol and Beyond

ORGANISATION FOR ECONOMIC CO-OPERATION AND DEVELOPMENT

ORGANISATION FOR ECONOMIC CO-OPERATION AND DEVELOPMENT

Pursuant to Article 1 of the Convention signed in Paris on 14th December 1960, and which came into force on 30th September 1961, the Organisation for Economic Co-operation and Development (OECD) shall promote policies designed:

- to achieve the highest sustainable economic growth and employment and a rising standard of living in Member countries, while maintaining financial stability, and thus to contribute to the development of the world economy;
- to contribute to sound economic expansion in Member as well as non-member countries in the process of economic development; and
- to contribute to the expansion of world trade on a multilateral, non-discriminatory basis in accordance with international obligations.

The original Member countries of the OECD are Austria, Belgium, Canada, Denmark, France, Germany, Greece, Iceland, Ireland, Italy, Luxembourg, the Netherlands, Norway, Portugal, Spain, Sweden, Switzerland, Turkey, the United Kingdom and the United States. The following countries became Members subsequently through accession at the dates indicated hereafter: Japan (28th April 1964), Finland (28th January 1969), Australia (7th June 1971), New Zealand (29th May 1973), Mexico (18th May 1994), the Czech Republic (21st December 1995), Hungary (7th May 1996), Poland (22nd November 1996) and Korea (12th December 1996). The Commission of the European Communities takes part in the work of the OECD (Article 13 of the OECD Convention).

Publié en français sous le titre :

Contre le changement climatique : bilan et perspectives du Protocole de Kyoto

Foreword

This report is the latest output of a decade-long involvement by the OECD in the economics of climate change. The global concern with climate change in 1997 led to the signing of the Kyoto Protocol, in which most OECD countries and some other countries committed themselves to reduce their emissions of greenhouse gases. The accumulation of these gases in the earth's atmosphere is assumed to have long-run implications for climate. The report discusses some of the main issues that need to be agreed for the Kyoto Protocol to be implemented. It also analyses the economic costs of reducing greenhouse gas emissions in line with the Kyoto commitments. Some main determinants of these costs are scrutinised, including the extent to which greenhouse gas abatement will be done where it is cheap. Realising that the Kyoto Protocol in itself will do little to avert climate change, the report then goes further to explore some of the requirements to establish a global agreement to stabilise atmospheric concentrations of greenhouse gases over the long run and the economic costs involved. The report also contains a tentative discussion of the costs of such climate change as may happen and what action can be taken to adapt as smoothly as possible.

The impetus to produce the report came from the OECD's Economic Policy Committee in the wake of the agreement on the Kyoto Protocol. A previous version of the report has been discussed by this committee and its Working Party No. 1, as well as the Environmental Policy Committee. The report is part of a wider effort by the OECD to address policy issues related to climate change. It is issued jointly with the report *National Climate Policies and the Kyoto Protocol*, which focuses on domestic policy actions to reduce greenhouse gas emissions. The OECD work on policies to address climate change is an important element of the three-year OECD project on Sustainable Development, initiated by OECD Ministers at their annual Council meeting in April 1998.

The main authors of the report were Jean-Marc Burniaux and Paul O'Brien with support from Christophe Complainville concerning modelling and other analytical aspects. Analytical inputs were also provided by Dong-Seok Choi. Anick Lotrous provided statistical support and secretarial assistance was provided by Muriel Duluc, Penelope El Ghadab and Jackie Gardel.

This report is published on my responsibility. The economics of climate change remains an area of many uncertainties and the report is therefore primarily aimed at stimulating discussion rather than providing final answers.

Donald J. Johnston
Secretary-General of the OECD

Table of Contents

Boxes

Tables

Figures

Taking Action Against Climate Change: the Kyoto Protocol

1. Introduction and summary

Meteorological evidence points to a rise in the average global temperature over the past century. The trend towards warming is consistent with predictions from global climate models, that higher concentrations of so-called greenhouse gases trap increasing amounts of heat in the earth's atmosphere. Increased concentrations of greenhouse gases result to a large extent from emissions arising from human activity – in particular the burning of fossil fuel. Against the background of increasing scientific consensus that man-made greenhouse gas emissions add to global warming, international efforts to reduce these emissions have thus far culminated in the Kyoto Protocol, agreed in December 1997.

The purpose of the current publication is fourfold, corresponding to its four main chapters. Chapter 2 describes some main aspects of the Kyoto Protocol and discusses some areas where choices in its implementation may have an impact on its success. The subsequent chapter quantifies the economic costs of implementing the Protocol under a variety of assumptions about how this will take place. Chapter 4 considers how, over the longer term, a credible effort towards stabilising concentrations of greenhouse gases can be made which includes participation by developing countries. The final chapter considers what kind of climate change is expected to occur over the next 50 to 100 years, which is largely independent of whether or not the Kyoto Protocol is successfully implemented, and discusses some aspects of the problems of adapting to such changes.

The Protocol contains targets for greenhouse gas emissions over the period 2008-12 for each so-called Annex 1 country; the countries involved include all OECD countries except Korea, Mexico and Turkey, and a number of economies in transition, most notably Russia. The targets are fairly ambitious in the sense that OECD emissions in 2008-12 will need to be some 20-40 per cent below the level they might be expected to reach without policy action. If countries were to respect their individual targets through domestic abatement measures, the cost to OECD countries may be a loss in annual real incomes of the order of magnitude of ¼-1 per cent provided that adjustment to higher energy prices is smooth. Given the size of energy price increases, this assumption cannot be taken for granted. The marginal abatement costs can be thought of in terms of a carbon tax that would typically correspond to a doubling or tripling of the international oil price from its 1995 level. Such large price changes will require substantial reallocation of resources, especially labour, between individual sectors of the economies. They will also lead to real wage losses which may be resisted with the end-result of higher unemployment. Preliminary analysis shows that in the latter case, costs of implementing the Kyoto targets could be substantially higher than the above estimates. The low estimated cost for OECD countries not only reflects the assumed flexible adjustment, but also that part of the cost will be borne by some non-OECD countries, as a result of the terms-of-trade changes induced by the lower demand for fossil fuel. When assessing the economic cost it has to be taken into account, however, that the model simulations do not include the "ancillary benefits" of emission reductions, such as reduced local pollution levels, which may also be significant. Finally, it is important to keep in mind that this assessment is based on econometric models which neglect the non-CO_2 GHGs and, therefore, tend to overestimate the economic costs of meeting the Kyoto targets.

The Protocol includes a number of mechanisms to reduce the cost of meeting the emission targets – and therefore increase the likelihood that the Protocol will be ratified and the targets met. The details of the Kyoto mechanisms, which include emission trading and the so-called Clean Development Mechanism, are not yet fully specified. It is important that the timetable, requiring agreement on details of the mechanisms by the end of year 2000, be met. The longer the delay, the later abatement measures are likely to be taken and, as illustrated by model simulations in Chapter 3, the higher the cost of meeting the emission targets is likely to be.

The costs of reaching the Kyoto targets can be significantly reduced by use of the flexibility mechanisms. Emission trading allows the marginal abatement costs in different Annex 1 countries to be equalised by trading emission permits so that emission reductions in one country, where they can be done relatively cheaply, may be counted against the target for another, where emission reductions are more costly. In the process, both countries can benefit. If trading succeeded in fully equalising marginal abatement costs across Annex 1 countries, the overall cost of meeting the targets could be cut by a third for OECD countries and become almost insignificant for all Annex 1 countries taken together, according to simulations with the OECD's GREEN model. Again, however, the strong assumptions behind and the uncertain nature of the model results should be borne in mind.

Emission trading raises a number of issues which need to be resolved for trade to deliver such efficiency gains. Some of these issues have a somewhat technical nature and relate to topics such as monitoring, verifying and enforcing compliance. The uncertainties in these areas have sometimes led to calls for restrictions on emission trading. Such restrictions, and the increased domestic action they imply, have also been justified with reference to the need for industrialised countries to show a good example which would induce developing countries to accept quantified emission limits too, as well as with reference to ancillary benefits of reduced domestic emissions (such as lower local pollution levels). The price of these restrictions would, however, be an increase in the costs of achieving the Kyoto targets. If, instead of general restrictions on trading, a subset of Annex I countries were to impose unilateral restrictions on emission trading, their costs could go up while those of other Annex I countries might fall.

Two specific implications of the participation by transition economies in full emission trading are worth mentioning. First, part of the apparent gains from trading would come from higher overall emissions: emissions in Russia and Ukraine are likely to be considerably lower than the Kyoto target even in the absence of any abatement measures; by allowing emissions by OECD countries to rise by that amount, trading reduces the latter's costs. Restricting trading in this so-called "hot air" might reduce the gains from trade by about a fifth, based on the amount of hot air present in the baseline scenario with the GREEN model. Such restrictions would, however, be politically difficult and also ineffective in reducing atmospheric concentrations of greenhouse gases over the longer term, because another provision ("banking") of the Protocol allows unused emission quotas to be carried forward for use (or trading) later. Second, with Russia and Ukraine potentially the dominant suppliers in the market for emission permits, there is a risk that the market may be imperfectly competitive; if collusion allowed Russia and Ukraine to keep permit prices significantly above the competitive level, the gains from trade could be reduced.

In addition to emission trading, the Kyoto Protocol contains two mechanisms for project-based emission transfers between countries. While so-called Joint Implementation among Annex I countries resembles emission trading in that it leads to transfers of emission rights, the Clean Development Mechanism allows Annex I countries to gain emission rights from investments to cut emissions by developing countries. The latter do not have emission targets and the transfers of emission rights therefore have to be made with reference to agreed, project-specific, baselines for emissions. This makes it hard to assess how important the Clean Development Mechanism will be. Preliminary analysis has assumed that only a fraction of the potential for low-cost abatement in developing countries will be exploited and suggests that the gains from the mechanism will therefore be modest if it is introduced in conjunction with full emission trading among all Annex I countries.

In the context of attempts to reduce global warming appreciably, the Kyoto Protocol is best seen as a starting point for a long-term effort towards a significant reduction in global greenhouse gas emissions, and eventually stabilisation or reduction of atmospheric concentrations. Chapter 4 shows that, although the science is far from sufficiently well-developed to give definitive answers, action by

Annex 1 countries alone will have only a marginal impact on concentrations, if emissions by non-Annex 1 countries grow as they are expected to without any constraint.

Inducing non-Annex 1 countries to accept such constraints is difficult, as evidenced by both the third and the fourth Conference of the Parties to the UN Framework Convention on Climate Change. This is not just because countries will be affected differently by climate change – some developing countries are likely to be more severely affected by it than most OECD countries – or because they have different valuations of these effects. Analysis in Chapter 4 suggests that the likelihood of a large number of countries voluntarily agreeing to cut emissions may be small in the absence of side payments. Furthermore, equity concerns may over the next half-century or so call for much larger cuts in emissions by Annex 1 countries than by developing countries, whose *per capita* emission levels are many times smaller. This, on the other hand, would conflict with concerns for global cost-minimisation, unless the flexibility mechanisms are sufficiently developed and implemented globally.

Chapter 4 considers a number of global emission scenarios which will move the world towards a stable concentration of greenhouse gases in the atmosphere. The scenarios differ both with respect to their ambition in terms of cutting global emissions and with respect to the distribution of the adjustment burden across countries. In particular, scenarios are constructed where emissions rights are distributed so as to reflect the economic disparities (thus taking into account the "ability to pay" of each country) or where countries will have to converge at the same, fairly low, level of emissions *per capita*. Under these two burden-sharing rules and given widely different abatement costs across countries, permit trading becomes a crucial instrument for keeping down costs. Depending on the degree of ambition with respect to emission paths, and assuming permit trading, annual costs at the world level could be of the order of ¼-1 per cent of GDP. However, revenues from permit selling may not provide sufficient incentives for all developing countries to participate in a global agreement, and Chapter 4 also considers the possibility that Annex 1 countries provide financial transfers to encourage participation. In this case, annual costs to Annex 1 countries could be sizeable if an ambitious emission reduction strategy is aimed for – the estimates range between 1½ and 2 per cent of Annex 1 GDP depending on the degree of ambition.

Despite the efforts being made in the context of the Kyoto Protocol and any likely follow-up agreement, some considerable further increase in atmospheric concentrations of greenhouse gases cannot be avoided. Chapter 5 notes that the resulting climate change is difficult to predict in the current state of knowledge, especially as regards precipitation patterns which, in many areas, will play a more critical role in determining the costs of climate change than changes in average temperatures. Few conclusions can therefore be drawn about the costs of climate change or about appropriate adaptation strategies in particular countries or for particular activities or sectors. Such studies of costs as have been undertaken appear to show relatively low costs over the next 50 to 100 years for most OECD countries. However, longer time horizons obviously matter – as do concerns for non-OECD countries, many of whom seem likely to bear much higher costs. Moreover, what is of interest is not just the mean expected outcome but also the variation around it – there may be a risk that beyond some concentration of greenhouse gases in the atmosphere, the associated climate change could be dramatic, with a substantial influence on the economy and society.

Whatever the change in climate may be, it is important that adaptation to it be as smooth as possible. To some extent, countries can learn from each other, given the wide differences that currently exist in climates. A number of countries also have some experience with adaptation, most prominently perhaps in the case of sea-level rise. In that area, because of externalities and perhaps also owing to economies of scale in information gathering, the public sector is typically involved in the necessary long-term planning and in resolving potential conflicts between different users of land. However, in many cases less rather than more public sector involvement will be necessary to achieve smooth adaptation – agricultural policy may be an example of this. At this stage it may be premature to draw very firm conclusions as to what type of policy change will be required to ensure smooth adaptation to climate change. Indeed, a general conclusion in the area of climate change is that there is a need for further knowledge. In particular, consideration of the accuracy of climate change forecasts, of the role of uncertainty and how fear of catastrophe should be valued (and indeed how a number of other aspects of climate change, such as species loss and other irreversible changes, should be valued) is required if the benefits of the Kyoto Protocol and its successors are to be usefully compared with their costs.

2. The Kyoto Protocol and the process of climate change

2.1. *Climate change*

Since the last substantial analytical work on climate change was presented to WP1 (summarised in OECD, 1995), the issue has lost none of its importance. Meteorological evidence from the 1990s has reinforced the view that some global warming is already under way. Scientific research finds that this tendency increasingly corroborates predictions from global climate models which incorporate the greenhouse mechanism, whereby certain gases produced by human activity – from industry, from households, from transport and from agriculture – trap increasing amounts of heat in the earth's atmosphere and add to the "natural" greenhouse effect.

Despite the considerable attention being paid to this problem over the past ten years and the declarations arising from the 1992 UN Framework Convention on

Climate Change (UNFCCC),[1] current indications are that emissions of greenhouse gases (GHGs) are rising faster than expected, that the initial UNFCCC target for emissions for the year 2000 will be substantially exceeded in almost all OECD countries and that the date by which stabilisation of world emissions (let alone the necessary reductions) can feasibly be expected is receding into the future.

There are still uncertainties over precisely how much additional warming the greenhouse effect will produce, on the consequences of this for climates and the associated effects on human activity and well-being. There also remain minority scientific views that global warming is either not happening or is largely unrelated to human activity, or is anyway beneficial on average. These do seem to be very much non-mainstream views, however; the 1995 report from the Intergovernmental Panel on Climate Change (IPCC)[2] concluded that "the balance of the evidence suggests that there is a discernible human influence on global climate".

This was the background against which the Parties to the UNFCCC met at their third conference (COP3), held in Kyoto in November-December 1997. The outcome of the meeting was the Kyoto Protocol, under which a number of countries have committed themselves to specific targets for emissions of greenhouse gases. However, many of its provisions left significant and essential detail to be agreed at subsequent Conferences of the Parties. COP4, in Buenos Aires in November 1998, made limited progress on these, but established a timetable for reaching agreement, with COP6, planned for late 2000, expected to settle any remaining implementation issues. The rest of this section reviews briefly the basic mechanisms through which human activity is thought to influence the climate and the next sections present some main features of the Kyoto Protocol.

The link between human activity and greenhouse gas emissions[3]

There are a large number of greenhouse gases (GHGs), *i.e.* gases whose presence in the atmosphere reduces the rate at which heat is radiated from the earth. After water vapour, overwhelmingly the most important greenhouse gas is carbon dioxide (CO_2). The Kyoto Protocol includes a further five gases or groups of gases: methane (CH_4), nitrous oxide (N_2O), hydrofluorocarbons, perfluorocarbons and sulphur hexafluoride.[4] The last three do not spontaneously exist in nature, while emissions of CO_2, CH_4 and N_2O may be of either human ("anthropogenic") or natural origin.[5] After CO_2, methane is the most important anthropogenic GHG. The main sources of CO_2 are fossil fuel burning and deforestation; large sources of methane are animal husbandry, rice production and natural gas venting. The fluorine compounds' origin in the atmosphere is largely leakage from their main uses in refrigeration and from the manufacture of aluminium and magnesium. Emissions of chlorofluorocarbons, also greenhouse gases, have declined rapidly since the

Montreal Protocol to phase them out came into force, and they are not covered in the Kyoto Protocol.

The link between emissions and concentration levels

In the short to medium term, and notwithstanding difficulties in establishing a precise relationship between emissions and concentration levels, the proportion of a given emission which remains in the atmosphere is thought to decline following a roughly exponential path over time as it is absorbed in the various "sinks" – the oceans, the soil, forests. The oceans absorb gases in solution, and also through sedimentation on the ocean floor; the rate of absorption from the atmosphere depends importantly on circulation within the oceans themselves, which affect the rate at which gases absorbed at the surface are transported to deeper water. Forests in equilibrium are not thought to be an important sink, since the amount of CO_2 they absorb through photosynthesis is balanced by respiration and decay of vegetation. They are important sinks while they are reaching maturity, however, and also if they are harvested for long-lasting wood products. Soils too may not serve as sinks when in equilibrium, but may absorb carbon when, for example, reverting from agricultural use to natural vegetation (as occurs in "set asides" for example). An area of uncertainty for forests and soils is the carbon fertilisation effect. The rate of photosynthesis, and hence carbon absorption by many plants, including young trees, is known to increase in laboratory conditions as atmospheric CO_2 concentration increases, provided the plant is not constrained by lack of nutrients or water, and provided the temperature is not above a certain (species specific) threshold. The extent to which this is true in nature is uncertain, especially for mature trees.

The rate at which different greenhouse gases are removed from the atmosphere varies: methane is almost entirely removed in a decade (partly through conversion into CO_2) or so, whereas carbon dioxide may persist for 200 years, and certain perfluorocarbons for many centuries. In principle, knowledge about the rates of exponential decay can be used to project concentration levels based on assumed emission paths.[6] Establishing such projections is very complex, however, taking into account what is known about sources and sinks, as well as atmospheric (and, increasingly, oceanic) circulation, and the range of uncertainty is fairly large; uncertainty relates not only to the mechanisms involved but also the measurement of the basic magnitudes.[7]

The determinants of climate change

Notwithstanding these uncertainties, the scientific consensus seems to be that human activity has contributed to an accelerating increase in the concentration of carbon dioxide in the atmosphere, and that this increase has already caused average global temperatures to be higher than they would otherwise have been.

Although over the last century records indicate that there have been periods when average global temperatures were falling (notably for about two decades after 1950), they do seem to have risen overall, and recently perhaps more rapidly than on average over the period; the melting of ice caps and retreating glaciers provide further evidence for this.

Increased concentrations have this effect because the GHGs absorb part of the long-wave radiation emitted by the earth's surface, reducing the proportion which escapes into space, thus increasing the equilibrium temperature required to maintain a balance between such outward radiation and incoming energy from the sun; provided that temperatures have been at (or below) this equilibrium level in recent history, actual temperatures can be expected to increase.

There is further uncertainty when it comes to interpreting the implications of temperature changes for the climate itself. Generally speaking, increased average temperatures are expected to lead to an intensified hydrological cycle – it will rain more. This is unlikely to be an evenly distributed increase, however; there may be increased frequency of both heavy rainfall in some places and droughts in others. In addition, average sea levels are likely to rise.

A principal force at work behind all climate patterns is the transfer of heat from the equatorial regions towards the poles. Higher overall temperatures can be expected to modify the associated atmospheric and ocean circulations. But detailed understanding of how these mechanisms work – especially ocean circulation – and their interaction with the nature of the land surface is relatively limited at the moment, so precise predictions are unavailable. Even if greenhouse gas concentration levels stabilise relatively early (which seems unlikely) most projections imply that climate change will continue for some time. The slow adjustment towards "equilibrium" makes it difficult to speak of the properties of the "new" climate associated with a given concentration level since it will be changing continuously through time. Hence, it is necessary to look very far ahead to get a full picture of the impact of changes – which, for a given region, could swing between positive and negative.

2.2. *Main features of the Kyoto Protocol*

The Protocol limits emissions from a group of countries known as "Annex I countries" (from the annex to the UNFCCC listing the original group, since expanded),[8] currently comprising all OECD countries (except Mexico) and a number of economies in transition.[9] Relative to their emissions in 1990, Annex I countries committed themselves to reduce their total GHG emissions (expressed in carbon equivalent tonnes),[10] by around 5 per cent on average for the period 2008-2012. There is some differentiation of the target reduction in emissions among Annex I countries. These range from 6 per cent in Japan, Canada, Hungary and Poland to 8 per cent in members of the European Union and some Eastern European countries. The commitment for the Russian Federation and the Ukraine, as well as New Zealand, is to emissions no higher than their 1990 levels. Some countries are even allowed to increase their emissions relative to 1990 levels.[11]

The Kyoto Protocol marked a retreat from one of the key initial aims of the UNFCCC – for the industrialised countries to limit greenhouse gas emissions in the year 2000 to no more than their level of 1990. By comparison with this target, the emissions objective is tightened but put back by about a decade: as mentioned, quantitative commitments refer to the average for the years 2008-12. The magnitude of the task now appears greater for most countries because of increases in emissions since 1990. As far as the OECD is concerned, only a few countries (the United Kingdom, Germany, Switzerland and Luxembourg) will meet the original target for the year 2000; most projections show that reaching the Kyoto targets for individual countries will require further policy action or making use of the flexibility mechanisms.

The flexibility mechanisms

Pursuing individual quantitative targets by domestic action only is a relatively costly way to reduce greenhouse gas emissions, since marginal costs of emission reduction may differ across countries as some have to take high-cost abatement measures while others do not fully exploit low-cost ones. The so-called flexibility mechanisms of the Kyoto Protocol allow a move towards alignment of marginal costs by providing ways in which emission reductions that take place in one country can be counted against another country's target.[12] Three mechanisms are available: emissions trading, joint implementation and the clean development mechanism.[13] As with some other aspects of the Protocol, the text is not precise on how the flexibility mechanisms should be implemented. The intention is that rules and procedures for these mechanisms will be developed by future COP (and its "subsidiary bodies") meetings; at COP4 it was decided that these aspects are to be finalised at COP6 in late 2000.[14]

- Emission trading

Article 17 of the Kyoto Protocol says that countries "may participate in emissions trading for the purposes of fulfilling their [2008-12 emissions target] commitments...". The intention is to allow Annex 1 countries which find it relatively easy to meet their Kyoto targets to reduce their emissions by more than is actually required, selling the surplus "permits" to countries whose abatement costs are higher.[15]

The relevant article of the Protocol states that emissions trading shall be "supplemental" to domestic action to reduce emissions. This reflected the desire of some of the participants to ensure that all countries took some measures to reduce domestic emissions and did not rely on buying "the right to pollute", but there is no consensus on what it will mean in practice. This, and other aspects of emissions trading, are taken up in more detail below.

- Joint Implementation (JI) and the Clean Development Mechanism (CDM)

The Joint Implementation mechanism (Article 6 of the Kyoto Protocol) allows one Annex I country to sponsor an emissions-reduction[16] project in another Annex I country and thereby acquire credit for the emission reduction, or part of it, as if it occurred in the sponsoring country, the same amount being deducted from the host country's allowed emission level. The important difference compared with direct trading of emission permits is the need to assess what emissions would have been without the project; the techniques for baseline definition are to be set by COP6. COP1 (1995) provided for an "Activities Implemented Jointly" (AIJ) scheme, and a number of pilot AIJ projects are already under way or planned.[17]

The Clean Development Mechanism (Article 12 of the Protocol) is similar to JI, with the difference – an important difference – that the host country be a non-Annex I country.[18] This difference could have important implications. In JI any error in the baseline in favour of either the host or the sponsoring country is automatically offset as far as global emissions are concerned, since both countries are subject to emission limits (and provided these limits are respected). In the case of the CDM, errors in baseline calculations may lead to higher global emissions than without the project since non-Annex I countries have no initial emission ceilings. Errors could also go the other way, leading to lower global emissions, but both sides will have an incentive to exaggerate the impact of the project in reducing emissions by inflating baseline estimates. Reflecting these differences in incentives, the Kyoto Protocol foresees the establishment of a board to supervise the operation of the CDM, whereas the basic requirement for a JI project to be accepted is that both countries agree to it.

JI and the CDM are similar to emissions trading in that their aim is to make abatement where it is cheap but there are also some important differences. Being project based they are in the nature of investments and intended also to promote technology transfer, whereas trading allows for "spot" transactions. JI and the CDM are likely to have rather higher transactions costs, since they require the calculation of a baseline and approval of the projects on a case by case basis by the countries concerned, and possibly an international board as well. Various approaches to solving the problem of establishing baselines are under discussion.[19]

Certified emissions reductions from CDM projects obtained between the year 2000 and 2008 "can be used to assist in achieving compliance in the first commitment period" (quoting Article 13). This is potentially quite an important provision, giving an incentive to reduce global emissions before the first commitment period (2008-12). The Protocol also implicitly requires that the emissions reduction be long lasting, but it is not clear how many years' worth of reductions can be credited to the sponsoring party. This and other aspects of the CDM mechanism remain to be clarified.

The CDM may have spillover effects on emissions trading, with consequences for the pattern of global emissions. Although non-Annex I countries have low *per capita* emissions, their emissions per unit of GDP are generally very high, and the cost of reducing their emissions intensity is low (relative to that in most OECD countries). The ease with which CDM projects can be approved and implemented will be crucial in determining their impact on emissions trading among Annex I countries: if the CDM were relatively friction free then the low cost of abatement in non-Annex I countries could put a low ceiling on the trading price of emissions permits, in turn meaning that actual reductions in emissions in Annex I countries would be relatively small (subject to supplementarity constraints, see below).

- Other aspects of the flexibility provisions

Supplementarity

Not all countries which signed up to the Kyoto Protocol view the "flexibility" measures in the same way. The text of the Protocol reflects a certain tension between *i)* the idea that emissions should be reduced in countries where they are already very high on a *per capita* basis and *ii)* the idea that they should be reduced at least cost in foregone output. The flexibility options result from *ii)*, but *i)* appears to be behind the introduction of the notion of "supplementarity".

The Articles defining emission trading and joint implementation carry wording on some kind of additionality condition.[20] Emission trading, for example, is supposed to be supplemental to domestic action to meet the Kyoto targets. At one extreme this could be taken to mean that domestic action should be almost sufficient to meet the targets before any trading is permitted, in which case trading would not only be supplemental, but also largely pointless; at the other extreme it could be interpreted to require just some minimal domestic action. On the basis of this provision, there are suggestions that trading should be limited, for example, to a certain proportion of total allowances.[21]

"Hot air"

Subsequent to the signing of the Protocol, a discussion has arisen over the issue of so-called "hot air". This refers to the possibility that some Annex I countries, notably some of the transition economies, might meet the Kyoto targets without any domestic action at all. In this case they would be able to sell their surplus emission allowance (the "hot air") without incurring any abatement cost. Predictions as to how much "hot air" there may be are highly dependent on forecast output in the former Soviet Union, particularly Russia and Ukraine. The present economic outlook for those countries suggests that there may be substantial amounts of "hot air". However, it is likely to be a relatively short-term phenomenon that will not persist into a second commitment period. Chapter 3 discusses the impact of "hot air" and possible restrictions on trading it for overall abatement costs.

Banking

Article 3 of the Protocol also provides for carrying over unused permits into future commitment periods. In general this is likely to be little used as a deliberate policy, given positive time preference and to the extent there are expectations that abatement costs will diminish over time. The banking provision is relevant to the "hot air" issue, since if countries are prevented from selling "hot air" they will automatically be able to carry it over into the future. Banning "hot air" trading would in this case do nothing to reduce cumulative emissions, only delay them somewhat – provided that assigned amounts for future commitment periods are not affected by actual emissions, which in practice they may well be given the sequential nature of negotiations.

Substitution between greenhouse gases

The Kyoto targets cover a "bundle" of gases. The target is expressed in terms of CO_2 equivalent, with the conversion factors for the different gases being given by their relative efficiency ("global warming potential") as a greenhouse gas.[22] This has implications for flexibility, since it potentially allows countries a wider choice of abatement strategies than those relating to CO_2 alone (provided their baseline emissions of the other gases are significant). As a result, a significant proportion of the necessary abatement could be achieved in the non-CO_2 gases, at lower cost than if CO_2 reductions were required for the full amount.[23] Measurement of non-CO_2 emissions is subject to more uncertainty than for CO_2, however, and there are suggestions that measurement problems are a further justification for limits on the proportion of a country's commitment that can be met by trading (see below).

Sinks

In addition to widening the scope of emissions targets to cover non-CO_2 gases, the Kyoto Protocol also introduces sinks into the process. Sinks are largely natural processes – not yet fully understood in many cases – and any anthropogenic component is generally due to large-scale land-use changes. The Protocol allows changes in take-up of greenhouse gases by sinks, certified as resulting from direct human-induced activities occurring after 1990, to be treated as a credit in the first commitment period. The main policy leading to such changes is reforestation. In some countries reforestation is already leading to significant increases in take-up of CO_2, but this will be of rather little importance to start with because most of this is due to measures that were taken before 1990. Moreover, the carbon sequestration potential in forests is initially relatively low but then increases (up to a point, as previously mentioned) as they mature, so the importance of sinks may increase in future commitment periods.

2.3. *Some implementation issues*

Among the flexibility mechanisms, much interest has been devoted to permit trading. As discussed in Chapter 2, permit trading has the potential to significantly reduce the economic costs of achieving the Kyoto targets. However, while the concept of emission trading is relatively simple in theory, there are many questions unanswered when it comes to setting up a trading system in practice. In particular, there are differing views on how compliance should be verified and enforced and who should be allowed to trade. This publication is not intended to recommend an appropriate structure for the trading system,[24] but, based on work carried out not least in the context of the Annex I Expert Group, this section highlights a number of issues that might affect trading and the overall aims of the Kyoto Protocol. Achieving early clarification on these issues will help implementing the Protocol in a cost-effective manner by creating a credible and predictable framework for forward-looking decisions by the private sector.

Ensuring compliance

Once in force, the Kyoto Protocol is a treaty commitment and governments will be obliged to ensure that it is respected. Nevertheless, elaboration of non-compliance responses is part of the mandate for COP6, and is important not least because of the inter-relation between compliance and permit trading: if compliance is not a constraint there is no incentive to buy permits; at the same time, some argue that the trading regime should be structured so as to encourage compliance.[25] In the United States sulphur dioxide trading regime, often used as a reference when discussing permit trading, the compliance problem is dealt with by a credible external sanction, well above the marginal abatement cost.[26] In principle, external sanctions could be imposed at an international level on countries exceeding their GHG emission limits, with trade policy sanctions, rather than fines, sometimes discussed as the instrument (though whether such measures would be consistent with WTO rules may raise certain issues). However, political consensus on this is unlikely at least at the moment.

Using the emission trading system to strengthen compliance could take the form of suspending a country's right to trade permits as a penalty for non-compliance. At first glance, the effectiveness of this sanction might seem limited: if a country sells a lot of permits and is then found to be emitting more than allowed, removing the right to sell permits that have been already sold is not a strong sanction. However, this will be a "repeated game" so the inability to sell permits in the future may, in fact, be quite a strong sanction in some cases – essentially for countries with substantially below average abatement costs, who could expect typically to be sellers. For countries likely to be net buyers of permits, the suspension of the right to buy permits might be an awkward penalty to impose for non-compliance: by the time non-compliance for the first commitment period can be established (likely to be late 2013 at the earliest) purchasing permits will be the only way to come into compliance. Here the repeated game aspect does not help much: if a country's non-compliance is deliberate then preventing it from buying permits in the future might have little impact in that country but would reduce the price of permits to sellers and tend to increase global emissions. Furthermore, countries have an incentive to attempt to manipulate measurement of emissions (which will be based on self-reporting plus expert review by the UNFCCC). In general, then, compliance enforcement will depend very much on individual countries' desire to meet treaty commitments rather than on threats of punitive action.[27]

Buyer or Issuer liability in permit trade?

Incentives for compliance are also involved in the discussion about buyer or issuer liability. The essence of the liability problem can be illustrated by an example in which country A has sold permits to country B. Which country will be held responsible if, at the end of the commitment period, A is found to have emitted more than its reduced stock of permits allowed, but less than its original stock?[28] Even where sanctions are not possible or not credible, liability rules can affect the incentives faced by different countries. Under Issuer[29] liability, any permits sold are sold irrevocably – in this example, B has nothing to fear. The wording of the Protocol establishes the liability of sellers to ensure they remain within their reduced emission limits. Under Buyer liability, the permits (or some of them) bought by B could be invalidated and it would be B that is treated as non-compliant (provided that the permits bought from A were indeed required for B to meet its obligations).

It is therefore argued that amending the Protocol to provide for Buyer liability would reduce the likelihood of this kind of non-compliance occurring, since the price of permits issued by potential non-compliers would be low, because of the risk that they will be unusable; hence countries wishing to sell permits would have an incentive to be compliant in order to increase their revenue, independently of whether they value the treaty commitment in itself. This presupposes that information about likely non-compliance is reasonably well available;[30] in practice, there will be very little information initially, which may discourage trading altogether until countries have a track record – not for some time. Until markets are deep enough,

Buyer liability may increase the likelihood of market segmentation, which would reduce the efficiency of permit trading in reducing overall abatement costs.

Buyer liability may also be unrealistic if countries wish to link domestic cap-and-trade systems with the international permit trading system, with domestic penalties for non-compliance. Suppose, for example, that a US enterprise purchases Russian permits and plans emissions on that basis, and some years later the permits turn out to be invalid because of excess emissions in Russia. Under Buyer liability the US enterprise is now in non-compliance and the US Government should penalise it. Theoretically, this should not be a problem because the US enterprise should have taken into account the probability that the permits would be invalid and planned its emissions in the light of the possible fine. In practice, however, this might be politically unacceptable, and could certainly generate quantities of litigation. Whether Buyer or Issuer liability[31] results in increased compliance will continue to depend on how seriously the different Parties take their Kyoto commitments.[32]

Entity trading?

Just as international trading should allow greenhouse gas emission abatement to be carried out in countries where it is least costly, domestic trading systems should reduce average abatement costs within countries. This would require governments allocating some part of the national quota of permits, or assigned amount, to plants, enterprises or other entities. If these domestic trading systems were linked to the international system the number of agents in the global market would increase and hence market power would tend to fall, which should improve the overall efficiency of trading (see Chapter 3 on the effects of market power).

It would not be possible to implement a domestic trading system covering all emissions of greenhouse gases, particularly those without point sources (such as methane but perhaps also household emissions of CO_2) where monitoring emissions is impossible and where there is not a fixed relationship between emissions and some measurable input or output. There would thus necessarily be a need for a mixture of policies, where some care would need to be taken to ensure that efficiency gains were not reduced by unnecessarily penalising or favouring particular kinds of emission.

The Kyoto Protocol does not expressly refer to entity trading (though it mentions legal entities or private entities in the context of JI or the CDM). However, even if there were opposition to entity trading under the Protocol itself, it would be relatively simple to link a domestic market to the international market: the government could regulate the supply of domestic permits as a function of its own holdings of international permits, operating on the domestic and international markets to keep the prices in line.[33] Allowing entities that are not themselves emitters to trade could also allow permit brokers to develop, potentially improving the functioning of the market.

Monitoring and verification

Another implementation problem, which has implications much beyond emission trading, is that emission volumes themselves can only be estimated with uncertainty. The situation varies among the different gases. Measurement of anthropogenic emissions of CO_2 is thought to be fairly accurate, because emissions arise largely from combustion of fossil and biomass fuels. The resulting production of CO_2 is closely related to the (known) carbon content of each fuel, and does not vary much according to combustion technology.[34] Provided use of such fuels can be measured, therefore, emissions can too.

Emissions of the other greenhouse gases are mostly related to industrial (for methane, mostly agricultural and waste disposal) processes or leakages from appliances either during use or on scrapping; emission rates vary, and vary considerably, according to the technology in use. Current estimates of emissions of these gases are subject to much greater relative margins of error than for CO_2. For some HFCs and SF_6 this may be up to a factor of 2 to 5; margins of error are lower for methane (+/-25 per cent, *i.e.* a factor of 1.25) and nitrous oxide (a factor of 2) but are still substantial (Gielen and Kram, 1998).

Another area where there is considerable measurement uncertainty is sinks. COP4 agreed that the adjustment to a Party's assigned amounts shall be equal to verifiable changes in carbon stocks during the period 2008-12 resulting from direct human-induced activities of afforestation, reforestation and deforestation since 1 January 1990. Unless direct measurement of absorption by sinks is attempted, itself likely to be subject to sampling error, it will have to be estimated on the basis of information on the relevant forestation measures, along with estimates of the rate of absorption by each different type of forestation.

Basic "default" methodologies for measurement have largely been agreed under the UNFCCC, with countries being encouraged to use better methods where possible. Through the process of "national communications" (which contain estimates of emissions) and expert review teams, measurement accuracy should be steadily improved (although the reviews do not directly check the accuracy of reported emissions figures). If the methodologies are inaccurate but unbiased then they may not lead to systemic excess emissions, unless the verification process is insufficient to detect possible "cheating". The incentive to cheat will be an increasing function of the sanction for non-compliance and the marginal abatement cost. But other factors, such as the capacity to comply, may also be important. Some have suggested that measurement inaccuracy justifies limiting trading in order to avoid excess emissions, or that participation in trading should be limited to countries whose data on emissions – at a national level and by entities that might be involved in trade – is reliable.

3. Implications of Kyoto

The focus of this chapter is on the macroeconomic costs of implementing the Kyoto Protocol. As described in the previous chapter, the Protocol as it stands leaves many questions open. Without a clear idea on how it will be implemented in practice, any assessment of the economic impact of the Protocol will have to be tentative. Quantifying the effects under different assumption may, however, allow an assessment of the contributions that various aspects of the Protocol, not least the various flexibility provisions, can give to the achievement of a cost-effective implementation.

The chapter focuses on near-term aspects of the Protocol; that is, until the end of the first commitment period (2008-12). It is based on results from a range of econometric models which all embody a number of simplifying assumptions – a prime simplification being the neglect of non-CO_2 GHGs. Among the model results considered are those obtained with the OECD's world general equilibrium model, GREEN (see Box 1).[35] The chapter first aims to identify the economic costs of the Protocol when countries and regions implement their emission targets individually – what is referred to as the "no-flexibility case" – and then analyses the impact of the so-called "flexibility mechanisms". Since it is unknown to what extent the flexibility mechanisms will be used, the approach is based on sensitivity analysis. Aspects of the Protocol lying beyond the first commitment period, including its impact on atmospheric CO_2 concentration, are taken up in Chapter 4.

3.1. *Economic costs of the Kyoto commitments: the no-flexibility case*

A first step in analysing the economic impact of the Kyoto Protocol is to consider the case where Annex 1 Parties meet their commitments individually. A scenario in which flexibility mechanisms play no role is a useful starting point for several reasons:[36] first, there is uncertainty about the extent to which these mechanisms will be allowed to work; second, there is little knowledge so far about some key technical aspects of some of these mechanisms and other aspects of flexibility in the Protocol – such as the potential, measurement and costs of sinks enhancement and reductions of non-CO_2 gases; third, such a scenario provides an upper bound estimate of the costs of applying the Protocol and can be used subsequently as a benchmark to evaluate the effectiveness of the flexibility mechanisms in reducing costs.

In all of the quantifications that have been reviewed, the required emission reductions have been fulfilled in terms of CO_2 alone – none of the models allow for non-CO_2 GHGs. Where emissions of other gases can be reduced at lower cost, the scenarios therefore may exaggerate the costs of implementing the Kyoto Protocol.

Box 1. The OECD's GREEN model

The quantified scenarios in this paper have been elaborated using the OECD Secretariat's GREEN model. GREEN is a multi-country, multi-sector, dynamic applied general equilibrium model which has been developed with the explicit aim of quantifying the economy-wide and global costs of policies to curb CO_2 emissions. In terms of the CO_2 issue, the model has a medium-term focus: it is simulated for the period up to 2050.

Because of the global nature of the GHG problem, specific attention is paid to the modelling of some key non-OECD regions. There are twelve detailed regional sub-models:

- four OECD regions (the United States, Japan, the EU and other OECD);
- eight non-OECD regions (the CIS, Eastern Europe, China, India, Energy-Exporting LDCs, Dynamic Asian Economies, Brazil and a Rest-of-the-World grouping).

GREEN has a simple recursive dynamic structure, in which saving decisions affect future economic outcomes through the accumulation of productive capital. In the version used here, capital accumulation is modelled in a putty/semi-putty fashion. Production sectors operate under constant returns to scale and markets are assumed to be perfectly competitive. Production and consumption are specified using systems of embodied Constant Elasticities of Substitution (CES) functions. Consumers maximise their utility and allocate consumption according to an Extended Linear Expenditure System.

There are eleven production sectors in GREEN, chosen to highlight the relationships between resource depletion, energy production, energy use and CO_2 emissions. Since the main source of manmade CO_2 emissions is the burning of fossil fuels, a key focus is on the energy sectors. Three sources of conventional fossil-fuel energy – oil, natural gas and coal – and one source of conventional non-fossil (so-called "carbon-free") energy are distinguished. The carbon-free energy source includes nuclear, solar and hydro power. Both carbon-based and carbon-free backstop technologies are available for all conventional energy sources at exogenously given costs and introduction dates.

The production side of each regional model describes in a detailed way the supply of fossil fuels and the use of fossil and non-fossil energy inputs in the production process. Some allowance is also made for shifts in the composition of production by treating agriculture as a separate sector, and by distinguishing between two broad aggregates, energy-intensive industries and other industries and services.

Consumer demand is split between four broad aggregates: food and beverages, fuel and power, transport and communication and other goods and services. Saving is treated implicitly as a "fifth good", and shifts in energy prices therefore affect both the structure of consumer demand and the consumption/saving mix through changes in real income.

The version of the model used in this paper (see Burniaux and Complainville, 1999) is updated and slightly modified compared to the one used in OECD's Model Comparisons Project (see Burniaux *et al.*, 1992 and Lee, Oliveira-Martins and van der Mensbrugghe, 1994). It is based on 1995 data elaborated by the GTAP project (Hertel, 1997).

On the other hand, there are also reasons to believe that these costs are underestimated, in particular because adjustment costs are not fully accounted for (see Box 3). The simulated impacts fall within a rather wide range, not surprisingly given the differences in the model structures and some ancillary assumptions. Moreover, the simulated impacts of policies to respect the Kyoto targets depend on the emission paths in the absence of such policies. These baseline scenarios – also referred to as Business-as-Usual (BaU) scenarios – differ between the various studies (the BaU with the GREEN model is described in Box 2). Nevertheless, despite these differences across studies, some insights appear robust.

Emission reductions are substantial

The Kyoto Protocol specifies the limitations on emissions by individual Parties in terms of 1990 emission levels. Expressed in this way, the Protocol implies what sounds like a rather modest reduction of about 5 per cent for Annex I countries as a whole. However, emissions have already risen relative to 1990 and will continue to grow before the Protocol is ratified and countries undertake effective abatement. Thus, as time passes, the abatement needed to reach the Kyoto targets tends to increase while the period over which it should be achieved becomes shorter. Table 3 shows the gap between emission growth during the period from 1990 to 1995 and the targets for Annex I countries. In most OECD countries, the gap between current emissions and target levels has increased substantially. This gap already exceeded 10 per cent by 1995 in Canada, Japan, the Untited States and a number of EU countries, including Denmark, Austria, Belgium and the Netherlands. Thus, for many OECD countries, it will take a very substantial effort to achieve the objectives of the Protocol unless large-scale reliance on the "flexibility mechanisms" is possible.

When measured relative to BaU scenarios, which differ across models but in all cases imply further growth in emissions, most OECD countries will have to cut their emissions by 20 to 40 per cent in 2010 (see Table 4, first column).[37] This will call for an effort of an unprecedented magnitude which is likely to require major structural adjustments. Russia and Ukraine (referred to as the CIS in the following)[38] are noticeable exceptions. Falling output, deregulation of energy markets and substantial subsidy reductions, have brought emissions in the CIS below their 1990 levels. In 2010, CIS unconstrained emissions could be substantially below commitments: the BaU scenario in GREEN suggests a gap ("hot air") of more than 14 per cent of the CIS commitment, some 130 million tons of carbon.

Marginal abatement costs are high in OECD countries

There are various ways to express the costs of reducing CO_2 emissions, the most straightforward being to calculate the marginal abatement costs for each Annex I Party. Table 4 shows that these marginal costs are likely to be rather high

Box 2. CO_2 emissions in the GREEN Business as Usual scenario

The first step in quantifying the economic impact of the Kyoto Protocol is to derive a plausible projection of future emissions as they would be expected in the absence of any policy action to restrain their growth – the so-called "Business-as-Usual (BaU) scenario. The emission targets set in Kyoto are expressed relative to Annex 1 countries' emission levels in 1990. Thus, the future growth of emissions, as projected in the BaU scenario, is critical in determining the efforts required to meet the Kyoto commitments.

Main assumptions behind the BaU scenario comprise future GDP and population growth rates. In addition, the BaU scenario incorporates assumptions about future technological developments in the field of energy production, including the future rate of Autonomous Energy Efficiency Improvement (AEEI) and the emergence of alternative energy options (referred to as "backstop" technologies).

Projected GDP growth

The model has been calibrated so as to reproduce given trends in GDP. These trends combine population growth as projected by the World Bank, with trends in productivity growth from various sources. Over the medium term, GDP growth rates are drawn from OECD *Economic Outlook* 63 and the associated Medium Term Projections. Longer-term assumptions about GDP growth are drawn from OECD Secretariat work on population ageing (see Turner *et al.*, 1998). Table 1 summarises GDP growth trends in the BaU scenario.

Energy related assumptions

Another important assumption is that each country/region has an annual autonomous energy efficiency improvement equal to 0.4 times its annual labour productivity growth. Moreover, the model considers three backstop energy sources: a carbon-based substitute to fossil fuels, a carbon-free substitute to fossil fuels and a carbon-free electricity source. These backstops are assumed to become available after 2010 at uniform prices across all regions. The prices have been set according to the Stanford Energy Modeling Forum (EMF) guidelines (7 333 1995 $ per terajoule for carbon-based fuel, 13 333 1995 $ per terajoule for the carbon-free fuel and 27 778 1995 $ per terajoule for carbon-free electricity generation). Finally, the current version of the model is based on the assumption that existing differences in energy prices across countries reflect distortionary taxes and subsidies which are assumed to remain constant. Subsidy rates have been updated on the basis of various recent studies, including World Bank (1997); OECD (1997*a*, *b* and *c*); Michaelis (1996). The international oil price is set to remain constant in real terms until 2000. Thereafter, it is endogenously determined, given the parameters of the reserve depletion function. These parameters have been calibrated on IEA medium-term projections and data from the World Energy Resources Program (USGS). As a result, the real world oil price increases by around 2 per cent annually on average over the period 1995-2050.

CO_2 emissions in the BaU

Projections of emissions depend on the above assumptions and their interactions through international energy prices. Thus, emissions are directly related to assumed future GDP growth. In addition, as oil and gas reserves are being depleted, energy demand shifts towards coal and the carbon-based fuel substitute. As the carbon content of these fuels is higher, this, in turn, tends to increase the amount of CO_2 emitted per unit of energy consumption. Table 2 shows the future emission trends as they are projected in the GREEN BaU scenario.

Box 2. **CO_2 emissions in the GREEN Business as Usual scenario** (*cont.*)

Table 1. **GDP projections underlying the baseline scenario with GREEN**
Average yearly growth rates, per cent

	1995-2000	2000-2005	2005-2010	2010-2030	2030-2050
United States	2.9	2.5	2.0	1.6	1.4
European Union	2.4	2.6	1.5	1.2	0.5
Japan	0.7	2.2	1.5	0.9	0.6
Other OECD	3.1	3.1	2.1	1.7	1.4
CIS	–2.5	3.5	4.5	4.0	3.5
Eastern Europe	3.6	4.6	4.1	3.6	3.1
Annex I	**2.1**	**2.6**	**1.8**	**1.5**	**1.2**
China	7.7	5.6	5.0	4.8	4.0
India	5.4	4.4	4.2	4.2	3.0
Dynamic Asian Economies	2.5	4.8	4.8	4.2	3.2
Brazil	1.7	3.1	2.9	2.8	2.0
Energy exporters	2.0	2.2	2.2	2.2	2.9
Rest of the World	2.7	3.2	3.0	3.0	2.9
Non-Annex I	**3.1**	**3.7**	**3.6**	**3.5**	**3.2**
World	**2.3**	**2.8**	**2.2**	**2.0**	**1.8**

Source: GREEN Model, OECD Secretariat.

Table 2. **CO_2 emissions in the baseline scenario with GREEN**
Average yearly growth rates, per cent

	1990-1995	1995-2000	2000-2005	2005-2010	2010-2030	2030-2050
United States	1.3	2.2	2.0	1.9	0.9	0.9
European Union	–0.3	1.4	1.4	0.9	0.6	0.5
Japan	1.6	1.7	1.7	1.5	–0.9	0.2
Other OECD	1.6	2.4	2.3	2.0	1.4	1.3
CIS	–7.4	–2.0	2.7	3.4	2.9	2.6
Eastern Europe	–3.1	1.2	1.8	2.0	2.6	2.3
Annex I	**–1.1**	**1.2**	**2.0**	**1.9**	**1.2**	**1.4**
China	4.8	5.5	5.5	5.2	5.2	3.1
India	6.4	3.5	3.8	3.8	3.9	3.0
Dynamic Asian Economies	8.8	3.0	3.3	3.2	2.2	2.5
Brazil	3.9	1.9	1.7	1.6	1.3	1.9
Energy exporters	3.8	1.9	2.2	2.2	2.2	2.1
Rest of the World	0.1	4.8	1.3	1.2	1.6	2.9
Non-Annex I	**4.6**	**3.9**	**3.8**	**3.7**	**3.9**	**2.9**
World	**0.8**	**2.2**	**2.7**	**2.7**	**2.6**	**2.3**

Source: GREEN Model, OECD Secretariat.

Table 3. **GHGs emission trends, Kyoto objectives and EU burden-sharing**

	Percentage change 1990-1995	Kyoto target for 2008-2012 (as % of 1990)
Non-EU OECD		
Australia	6	8
Canada	10	–6
Czech Republic	–24	–8
Hungary	–24	–6
Iceland	5	1
Japan	8	–6
New Zealand	0	0
Norway	6	1
Poland	–22	–6
Switzerland	–2	–8
United States	5	–7
European Union (EU)		**–8.0**
Burden-sharing targets:		
Austria	1	–13.0
Belgium	6	–7.5
Denmark	10	–21.0
Finland	3	0.0
France	0	0.0
Germany	–12	–21.0
Greece	6	25.0
Ireland	4	13.0
Italy	2	–6.5
Luxembourg	–24	–28.0
Netherlands	8	–6.0
Portugal	6	27.0
Spain	2	15.0
Sweden	3	4.0
United Kingdom	–9	–12.5

Source: UNFCCC "official" national data.

for OECD countries if they have to meet their targets individually. A majority of models reports marginal costs ranging from 100 to around 300 1995 dollars per ton of carbon or even more in Japan.[39] Though at variance with the results from GREEN, most available studies point to higher marginal abatement costs in the European Union, Japan and the rest of the OECD region (grouping mainly Canada, Australia and New Zealand), than in the United States.[40] In contrast, marginal costs are lower in Eastern Europe and zero in the CIS, where the limit decided in Kyoto is not a binding constraint.

Total economic costs seem low if real wages are flexible

The right-hand column in Table 4 indicates that, according to most models, the total economic costs – expressed as percentage of GDP or total real income – would

Table 4. **Estimates of the economic costs of implementing the Kyoto Protocol in 2010 without the flexibility mechanisms**

		Emission change (in per cent) relative to the BaU in 2010	Marginal cost in 1995 $ per ton of carbon	Total cost as percentage reduction of GDP in 2010
United States	WorldScan[1]	–28	41	–0.2
	G-Cubed[2]	–39	82	–0.4
	POLES[4]	–23	90	–0.2
	RICE[14]	–27	148	–0.9
	AIM[5]	–25	170	–0.4
	CETA[13]	–34	191	–1.9
	SGM[12]	–43	211	..
	MIT-EPPA[11]	–39	216	..
	GREEN	**–36**	**231**	**–0.3 –0.3 (HEV)[16]**
	MS-MRT[10]	–41	272	–1.8
	Merge[3]	–42	304	–1.0
	GTEM[8]	–35	365	–2.1
	Oxford Model[9]	–42	464	–2.2
Western Europe	WorldScan[1]	–29	83	–0.3
	PRIMES[6]	–13	72	..
	POLES[4]	–22	141	–0.1
	RICE[14]	–24	184	–0.5
	GREEN	**–22**	**189**	**–0.2 –0.8 (HEV)[16]**
	MS-MRT[10]	–21	200	–0.6
	GEM-E3[7]	–15	218	..
	AIM[5]	–20	226	–0.4
	G-Cubed[2]	–44	253	–1.7
	MIT-EPPA[11]	–35	319	..
	SGM[12]	–21	452	..
	GTEM[8]	–32	760	–1.0
	Oxford Model[9]	–25	1 077	–1.9
Japan	WorldScan[1]	–22	93	–0.8
	G-Cubed[2]	–36	106	–0.5
	GREEN	**–32**	**182**	**0 –0.2 (HEV)[16]**
	POLES[4]	–27	260	–0.3
	AIM[5]	–26	266	–0.2
	RICE[14]	–36	279	–0.8
	GRAPE[15]	–82	356	–0.2
	SGM[12]	–40	411	..
	MS-MRT[10]	–30	452	–1.7
	Merge[3]	–33	559	–0.8
	MIT-EPPA[11]	–39	576	..
	GTEM[8]	–26	733	–0.8
	Oxford Model[9]	–16	1 213	–1.9
Eastern Europe and the CIS				
Eastern Europe	WorldScan[1]	–10	4	0.0
	GREEN	**–15**	**32**	**–0.3 0.1 (HEV)[16]**
	GTEM[8]	–24	43	–0.5
CIS	WorldScan[1]	0	2	–1.1
	GREEN	**4**	**0**	**–0.1 –1.7 (HEV)[16]**
	AIM[5]	1	0	–0.2
	GTEM[8]	2	0	0.0

Table 4. **Estimates of the economic costs of implementing the Kyoto Protocol in 2010 without the flexibility mechanisms** (*cont.*)

1. Emission reductions and economic costs are from Gielen and Koopmans (1998); carbon taxes are from Bollen, *et al.* (1998).
2. Global General Equilibrium Growth Model, Australian National University, University of Texas and US Environment Protection Agency; derived on the basis of Weyant and Hill (1999), Figures 7, 8 and 9.
3. Model for Evaluating Regional and Global Effects of GHG Reductions Policies, Stanford University and Electric Power Research Institute; derived on the basis of Weyant and Hill (1999), Figures 7, 8 and 9.
4. Figures from POLES are derived on the basis of Figure 2, p. 35 and Figure 3, p. 36 of Capros in OECD (1998*a*).
5. Asian Pacific Integrated Model, National Institute for Environmental Studies (NIES–Japan) and Kyoto University; derived on the basis of Weyant and Hill (1999), Figures 7, 8 and 9. Figures for the CIS are quoted from Kainuma, Matsuoka and Morita in OECD (1998*a*), P. 167.
6. Figures from PRIMES refer to EU8 and are derived on the basis of Figure 4, p. 41 of Capros in OECD (1998*a*).
7. Figures from GEM–E3 refer to EU14 and are derived on the basis of Figure 5, p. 43 of Capros in OECD (1998*a*).
8. Global Trade and Environment Model, Australian Bureau of Agriculture and Resource Economics (ABARE, Australia); derived on the basis of Weyant and Hill (1999), Figures 7, 8 and 9. Figures for the CIS and Eastern Europe are quoted from Tulpulé *et al.* in OECD (1998*a*).
9. Oxford Economic Forecasting model; derived on the basis of Weyant and Hill (1999), Figures 7, 8 and 9.
10. Multi–Sector, Multi–Region Trade Model, Charles River Associates and University of Colorado; derived on the basis of Weyant and Hill (1999), Figures 7, 8 and 9.
11. Emissions Projections and Policy Analysis Model, Massachusetts Institute of Technology; derived on the basis of Weyant and Hill (1999), Figures 7, 8 and 9.
12. Second Generation Model, Batelle Pacific Northwest National Laboratory; derived on the basis of Weyant and Hill (1999), Figures 7, 8 and 9.
13. Carbon Emissions Trajectory Assessment Model, Electric Power Research Institute and Teisberg Associates; derived on the basis of Weyant and Hill (1999), Figures 7, 8 and 9.
14. Regional Integrated Climate and Economy Model, Yale University; derived on the basis of Weyant and Hill (1999), Figures 7, 8 and 9.
15. Global Relationship Assessment to Protect the Environment Model, Institute for Applied Energy (Japan), Research Institute of Innovative Technology for Earth (Japan) and University of Tokyo; derived on the basis of Weyant and Hill (1999), Figures 7, 8 and 9.
16. HEV = equivalent-variation of household real income.

be around 1 per cent or below.[41] Although not imposing any carbon limitation,[42] the CIS would lose from an adverse terms-of-trade effect due to reduced energy export revenues. Conversely, the slight real income gain in Eastern Europe results from a drop of the world oil price.

These global cost estimates may appear surprisingly modest in view of the size of emission cuts and their high marginal costs. When interpreting these results it should be borne in mind that global economic models (like GREEN, Merge or WorldScan) provide cost indications which correspond roughly to the "dead-weight losses". Thus, they consider the efficiency losses (in terms of producer and consumer surplus) which are associated with the resource reallocation resulting from a carbon limitation rather than the total abatement cost (as measured, for example, by the revenues of the carbon tax required to reduce emissions). The simulations also assume that domestic implementation policies are cost-effective in the sense of raising marginal costs of CO_2 emissions by the same amount across sectors (even if pre-existing distortions are maintained, see page 59). Hence, these low-cost estimates should not hide the fact that some sectors and some countries, including among non-Annex 1 Parties, would be substantially affected. Moreover, as discussed later in the paper, the impact of these emission reductions on atmospheric concentrations and climate change will be small. To have a significant impact on the global warming process will require much greater efforts, involving a larger number of countries, and so the costs (and the gains to minimising them) will be correspondingly greater.

Furthermore, most global economic models tend to underestimate the economic costs of carbon limitations, especially over the short and medium term, because they assume labour and capital to be reallocated smoothly in response to higher carbon prices. Carbon abatements as those implied by the Kyoto Protocol are likely to hurt some sectors very strongly, causing existing capacities to become unprofitable and labour forces to shrink. Abatement costs, by raising the cost of living, may also generate unemployment depending on the degree of rigidity of real wages over the medium term. These costs have been little analysed in the literature so far.[43]

The GREEN model contains some adjustment costs associated with capital turn over.[44] But, as the results in Table 4 show, even with these costs included aggregate cost estimates remain modest. As for the amount of labour reallocation induced by the emission reductions, estimates using the GREEN model may be on the low side but suggest that it could be as low as 0.2 per cent of the total labour force in 2010.[45] The transitory costs associated with this reallocation of labour should not cause a substantial increase of the aggregate cost estimate – especially when taking into account that annual job turnover rates lie in the range of 15 to 30 per cent in most OECD countries (OECD, 1996). By contrast, Box 3 illustrates that the existence of real wage rigidities may dramatically increase the aggregate cost of meeting the Kyoto targets. A number of studies[46] have shown that, at least over the short term, real wages fail to adjust so as to offset unemployment changes. However, quantifying these rigidities econometrically remains difficult and the resulting estimates are not easily introduced in a general equilibrium framework. So the approach followed in Box 3 is primarily illustrative. It consists in testing alternative but very crude assumptions about real wage rigidities in order to quantify how they could affect the cost of implementing the Kyoto Protocol. The results indicate that, depending on the type and extent of rigidity, aggregate costs may be amplified several fold. Another way of stating the same conclusion is that flexible and adaptable labour markets will do a lot to keep down the aggregate economic costs of implementing the Protocol.

It should be noted, however, that there are also arguments which point in the direction of economic costs being lower than simulated. First, the revenues from carbon taxes or permit sales may be used to reduce taxation which currently distorts resource allocation. Second, and as previously mentioned, to the extent emissions of other gases than CO_2 can be reduced more cheaply, the costs will be lower (see page 61). Third, the reduction of other distortions in the economy, such as energy subsidies (see page 59), makes it possible to meet the Kyoto targets at a lower cost.

Box 3. The effects of real wage rigidities

Most model simulations of the Kyoto Protocol, including those of the GREEN model, are based on the assumption that wages adjust flexibly so that there is no change of unemployment.[1] However, casual observation backed up by both theory and econometric studies suggests that labour markets of many OECD countries are characterised by a significant degree of real wage rigidity over the medium term. At best, therefore, results from these model scenarios have to be interpreted as effects occurring over the longer run. Thus, most scenarios may underestimate the costs of achieving the Kyoto targets, at least over the medium term, and the purpose of this box is to provide a benchmark evaluation of this potential underestimation.

OECD (1999*a*) indicates that the speed of adjustment of real wages towards their long-run equilibrium level is much lower in the European Union than in the United States, with only half of the adjustment being completed over a period of four to five years. A range of similar results is available in the literature. However, econometric estimates of the link between real wages and unemployment are uncertain. Moreover, these estimates do not easily translate into a general equilibrium model which considers only relative price changes but not inflation. Therefore, the approach followed here is more crude. Basically, it consists in making the extreme assumption that real wages are fully rigid in OECD countries so as to provide an upper-bound benchmark of the potential impact of wage rigidities on the costs of implementing the Kyoto Protocol.

Scenarios with flexible wages – as simulated with the standard GREEN model – indicate that, as emission abatements raise the cost of living, the equilibrium real wage falls. With no permit trading, real wages fall by between 3 and almost 4 per cent by 2010 in OECD countries. However, wage earners are partly compensated for their loss of purchasing power by receiving parts of the revenues from carbon taxes or permit sales.[2] Taking these refunds into account, wage earners would still lose 0.5 to 1.5 per cent of their "disposable real wage"[3] in 2010. The subsequent analysis considers the alternative assumption that the real wage is fully rigid – *i.e.* set exactly equal to its level in the BaU scenario. As the equilibrium constraint in the labour market is relaxed, the model endogenously generates unemployment if the fixed real (or disposable real) wage exceeds the corresponding equilibrium real wage.

Table 5 demonstrates that real wage rigidities may potentially increase the costs of meeting the Kyoto targets very substantially. With countries pursuing their abatement targets individually (the upper panel in Table 5) together with the most extreme assumption that real wages are fully rigid, GDP in OECD countries could fall by 4 per cent in 2010 (against 0.2 per cent with flexible wages) and household real income by 4.6 per cent (against 0.5 per cent with flexible wages). The impact is noticeably weaker in Japan (real income falling by 1.6 per cent; GDP by 1.4 per cent) in line with the smaller reduction of the equilibrium real wage following the implementation of the Kyoto constraint in the standard simulation. The effects on unemployment are dramatic: for the OECD on average, the unemployment rate could rise by some 5 percentage points. The simulations implicitly assume the labour supply to remain constant. In practice, a significant drop in employment would probably be associated with a fall in labour supply, partly cushioning unemployment.

Box 3. **The effects of real wage rigidities** (*cont.*)

The assumption of rigid "disposable real wages" leads to somewhat lower aggregate costs, although still significantly higher than the corresponding estimates with flexible wages (right-hand panels of Table 5). In this scenario, it is assumed that wage earners, being partly compensated for their loss of real purchasing power by part of the carbon revenues, would take these compensatory payments into account in formulating their wage claims, therefore implying a lower fixed real wage than in the previous scenario. Therefore, the costs in this scenario are lower than with fully rigid real wages (especially in the United States where carbon revenues/ proceeds are relatively more important) but still remain substantial in some countries/ regions, such as the EU (over 3 per cent in 2010). For OECD countries on average, the GDP loss would amount to 1.4 per cent in 2010 (against 0.2 per cent with flexible wages) and the real income loss to 1.8 per cent (against 0.5 per cent with flexible wages), with the EU being proportionally much more affected than the other OECD countries. The unemployment consequences would also be correspondingly smaller with this scenario.

The bottom panel of Table 5 also reports the corresponding costs with permit trading (permit trading is discussed in page 37). With fully rigid real wages, aggregate costs are roughly cut by half. With rigid "disposable"real wages, the aggregate cost saving achieved by trading permits is less important.

1. Two studies, at least, have taken some kind of wage rigidity into account in estimating the macroeconomic costs of Kyoto. Bayar (1999) analyses the effect of CO_2/energy taxes on European unemployment under the assumption that wages are set in a bargaining process. But the study does not identify the impact of this particular assumption on costs which, in any case, is offset by the effect of recycling the tax revenues through reductions in the social security contributions. McKibbin *et al.* (1999) incorporate nominal wage rigidities in their G-Cubed model but, again, the impact of this assumption on cost estimates is not presented. The results of the Oxford Economic Forecasting Model reported in Table 4 suggest that adjustment costs may potentially increase the estimates of the economic losses incurred by complying to the Kyoto targets.
2. In GREEN, revenues from carbon taxes or permits sales are redistributed as lump-sum transfers to households so as to maintain the government net balance unchanged. In practice, however, not all GHG abatement is likely to generate government revenues and, therefore, revenue re-cycling will be less than assumed in these simulations.
3. The "disposable" real wage is defined as the real wage augmented with the part of the carbon revenues allocated to wage earners. In the simulations for this box, the revenues are partly allocated to unemployed (assuming a wage replacement rate of 30 per cent) and partly distributed between wage earners and capital owners in proportion to their income shares in total GDP.

Higher prices and lower output of energy

Carbon taxes or prices of domestically traded permits in the range corresponding to the marginal costs reported in Table 4 are likely to generate sharp increases

Table 5. **Economic impact of real wage rigidities in the context of the Kyoto Protocol**

Per cent deviation relative to BaU in 2010

	Fully flexible real wage		Fully rigid real wage			Rigid "disposable" real wage[1]		
	GDP	Household real income	GDP	Household real income	Unemployment rate	GDP	Household real income	Unemployment rate
With no permit trading:								
Unites States	–0.3	–0.3	–4.6	–5.1	5.8	–0.7	–0.7	0.6
European Union	–0.2	–0.9	–4.8	–6.1	5.8	–3.1	–4.0	3.7
Japan	–0.0	–0.2	–1.4	–1.6	2.0	–0.3	–0.5	0.4
Other OCDE	–0.3	–0.7	–4.7	–5.7	6.1	–1.2	–1.6	1.3
CIS	–0.3	–1.7	–0.3	–1.9	0.0	–0.3	–1.8	0.0
Eastern Europe	–0.2	0.1	–0.3	–0.5	0.0	–0.3	–0.3	0.0
Total OECD	–0.2	–0.5	–4.0	–4.6	5.1	–1.4	–1.8	1.7
Total Annex I	–0.2	–0.5	–3.8	–4.5	4.9	–1.4	–1.8	1.7
With permit trading:								
United States	–0.2	–0.4	–2.4	–2.9	3.0	–0.7	–1.0	0.8
European Union	–0.1	–0.4	–2.4	–3.0	2.9	–1.6	–2.0	2.0
Japan	–0.0	–0.2	–0.8	–0.9	1.1	–0.3	–0.4	0.4
Other OCDE	–0.2	–0.6	–2.4	–3.2	3.1	–0.9	–1.4	1.1
CIS	–1.1	8.5	–1.1	7.5	0.0	–1.1	8.1	0.0
Eastern Europe	–0.5	1.1	–0.5	0.6	0.0	–0.5	0.8	0.0
Total OECD	–0.1	–0.4	–2.1	–2.5	2.6	–0.9	–1.2	1.2
Total Annex I	–0.1	–0.1	–2.0	–2.2	2.5	–0.9	–1.0	1.1

1. In the sense that the "disposable" real wage cannot exceed its BaU level although, in practice, it has been set at a slightly lower level than in the BaU (–0.2 per cent).

Source: GREEN Model, OECD Secretariat.

in energy product prices. Figure 1 reports the price increases simulated by the OECD GREEN model. The coal price in OECD countries would increase by 200 to 400 per cent (relative to the BaU scenario). Other energy prices would increase too: the average price consumers pay for energy services would increase by more than 50 per cent (except in Japan).[47] This substantial impact on consumers is partly offset by the assumption in the GREEN model that additional revenues from carbon taxation are redistributed to consumers in a lump-sum way.[48]

The output of energy producing sectors is reduced substantially in 2010 (Figure 2), with coal output falling most, while the production of electricity is either reduced by much less or even increases. There is widespread concern that energy-intensive industries (steel, cement, paper and pulp, *etc.*) could be adversely affected by the Protocol. This is an area where results differ somewhat. According to simulations with the Merge model, production in these sectors in some Annex I Parties could fall by as much as 10 per cent in 2010 (Manne and Richels, 1998). Results from GREEN suggest smaller impacts, with the production of energy-intensive sectors cut by 1 per cent only in Japan and the European Union, 3 per cent in the United States and 4 per cent in other OECD countries.

Oil prices will be hit but carbon leakage may not be a concern

Implementation of the Protocol is likely to have a large impact on the international market for crude oil. According to results with GREEN, reduced oil consumption by Annex I countries causes a real income loss of 3 per cent among the oil-exporting countries, proportionately much higher than any loss estimate for Annex I Parties. By contrast, some oil-importing developing countries (for instance, Brazil) may benefit from lower oil prices.

Such lower oil prices are likely to lead to "carbon leakage" whereby part of the emissions reduction achieved in Annex 1 countries could be offset by additional emissions in non-Annex 1 countries. This may occur through migration of energy intensive industries away from Annex 1 countries where they will become less competitive, though GREEN simulations suggest rather small reductions in such output, in the range of 1 to 5 per cent (Figure 2), and thus a small carbon leakage effect from this source. More generally, carbon leakage will result from substitution towards cheaper energy in all non-participating countries.

Carbon leakage may increase with the degree of international capital mobility, but model results indicate that this factor has little impact; it could even contribute to reduce leakages under certain circumstances (McKibbin *et al.*, 1997; Burniaux and Oliveira Martins, 1999). The existence of economies of scale and imperfect competition, leading to international relocation of firms, are other factors which could potentially determine the magnitude of leakages, although their influence has not been quantified so far. On the whole, there is widespread disagreement among

Figure 1. **Implementing the Protocol: energy price increases in 2010 (without permit trading)**
(per cent)

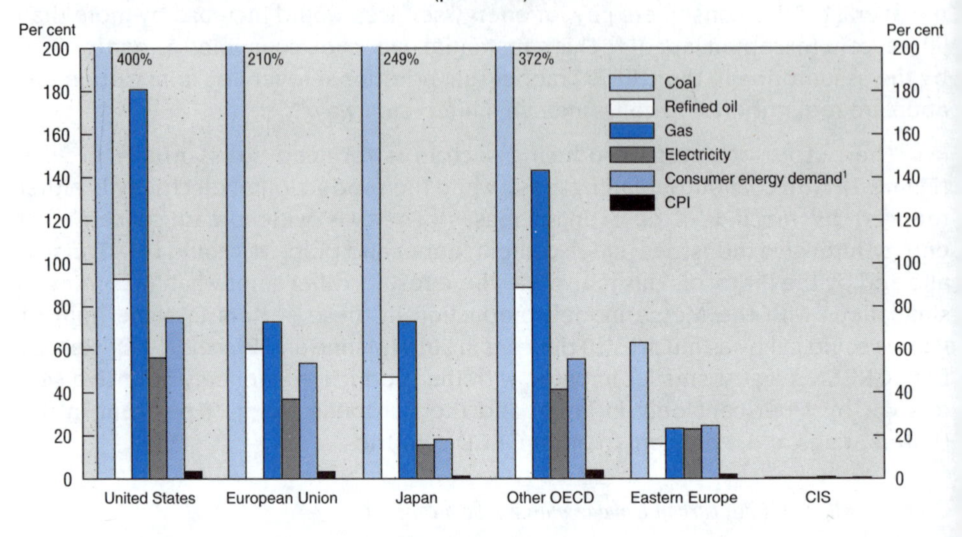

Figure 2. **Implementing the Protocol: output changes in 2010 (without permit trading)**
Per cent

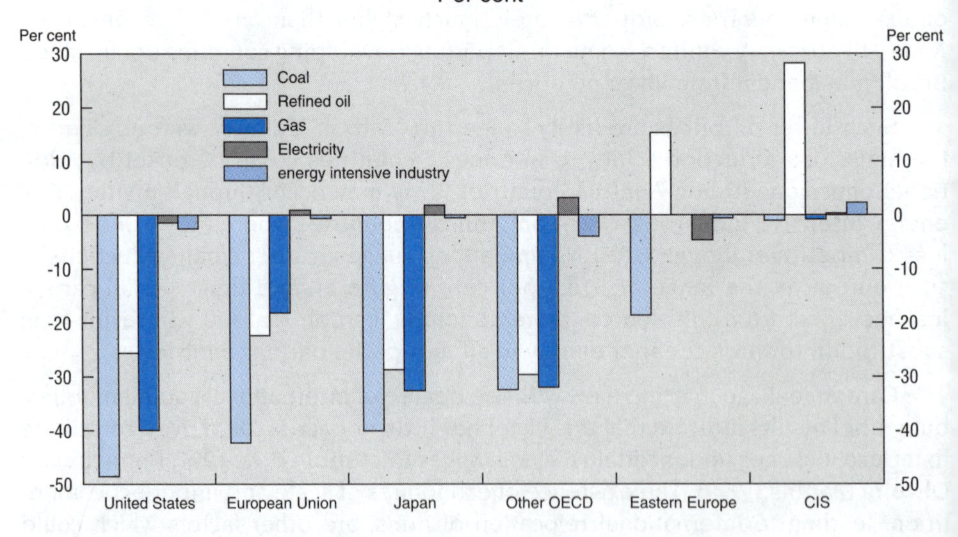

1. Average price of the "fuel and power" consumer aggregate of GREEN.
Source: GREEN Model, OECD Secretariat.

available models about the size of carbon leakages that would occur following the implementation of the Kyoto Protocol (see Box 4). In simulating the Kyoto Protocol, the Merge model (Manne and Richels, 1998) the G-Cubed model (McKibbin *et al.*, 1998) and the MIT-GTAP model (Babiker and Jacoby, 1999) show leakage rates (calculated as the ratio of additional emissions in non-Annex 1 countries to the total emissions reduction in Annex 1 countries) below 10 per cent, whereas the GREEN model estimates leakage rates around 5 per cent.[49] In contrast, other models – such as, for instance, the model WorldScan (Bollen *et al.*, 1999) and the MS-MRT model (Bernstein *et al.*, 1999) report much higher leakage rates, approaching 20 per cent of the initial emission reduction in the Annex 1 countries. These divergences are not well understood but the analysis summarised in Box 4 suggests that they could reflect, at least in part, different assumptions about the supply responses of coal and oil producers.

3.2. *Reducing the costs of Kyoto: making use of the flexibility mechanisms*

The scenario analysed in the previous section pointed to large differences of marginal abatement costs across Annex 1 Parties. This implies that the Protocol's provisions for flexible, market-based instruments may significantly reduce the cost of achieving the overall emissions reduction target. Drawing on Chapter 2, this section reviews these flexibility instruments and attempts to quantify their potential to reduce the economic costs.

The mechanisms considered in the Protocol can be characterised in terms of three basic categories of flexibility:[50] *i)* the "where" flexibility which allows emission reductions to be redistributed across Parties so as to minimise costs; *ii)* the "when" flexibility which permits the reallocation of emissions reductions over time; *iii)* the "what" flexibility which allows a choice between cutting six different gases and/or enhancing sinks instead of cutting emissions. The Protocol also leaves Parties free to choose how to implement their limitation domestically, for instance by applying taxes or imposing controls. As part of this "how" flexibility, this section also analyses the possibility that some Parties remove remaining subsidies in their energy markets.

"Where" flexibility

- "Where" flexibility among Annex 1 Parties

Existing studies agree that the potential for cost-saving by emission trading (ET) and joint implementation projects (JI) is very substantial.[51] For instance, the Stanford Energy Modeling Forum (EMF), by comparing the results from six world-wide models, has estimated that international trading would reduce the

Box 4. Carbon leakage: an unresolved issue

Unilateral abatements – such as by Annex I countries under the Kyoto Protocol – run the risk of being offset, at least partially, by so-called "carbon leakage". Carbon leakage refers to possible increases of carbon emissions in non-Annex I countries in response to the emission reductions implemented by Annex I countries. High carbon leakages would reduce the environmental impact of the Protocol while making its extension to non-Annex I countries even more problematic (see Box 6).

Leakage results from complex spillover effects within and between energy and non-energy markets. In the absence of direct empirical evidence on the magnitude of carbon leakages, the only way to assess their potential importance is to rely on model simulations. However, existing global models have failed so far to provide a coherent view on the magnitude and the distribution of the leakages generated by the implementation of the Protocol. In the absence of general agreement on the size of the leakage effect, this Box discusses some of the main determinants of leakage.

Underlying mechanisms of carbon leakages

Carbon leakages can be generated through different mechanisms. For simplicity, it is convenient to distinguish two main channels:

- The first operates via non-energy markets. Carbon abatements imposed unilaterally raise production costs which affect the competitiveness of energy-intensive industries. These industries lose market shares on the international markets to the benefit of industries located in countries that do not reduce their emissions, causing a corresponding shift of the production of energy-intensive goods. Models usually represent these mechanisms by way of trade substitution elasticities (the so-called Armington elasticities). The induced reallocation of foreign direct investment may contribute to enhance or to mitigate the carbon leakages occurring through the non-energy channel.
- The second channel relates to international energy markets. Unilateral carbon abatements in a group of countries corresponding to a large fraction of world carbon demand would cause a fall of the international price of carbon, thus increasing energy demand and carbon emissions in the rest of the world. The structure of the international carbon market clearly matters here.

Although oil can be considered a homogenous good, there is more uncertainty about the degree of integration of the world coal market.[1] The supply response of the carbon producers – especially coal producers – will also be influential, perhaps even more so than the structure of the international carbon market. Indeed, the potential for reducing world carbon emissions ultimately relies on the decision by the carbon producers to keep extracting carbon or to leave it in the ground. The crucial parameters that characterise the supply behaviour of carbon producers are the supply elasticities for coal, oil and natural gas.

Box 4. Carbon leakage: an unresolved issue (*cont.*)

Other factors may also prove important in determining the amount of leakage likely in the context of the Kyoto Protocol:

- The existence of "hot air" in the Russian Federation and Ukraine implies that emissions in these two countries are not subject to any binding constraint. This raises the possibility of carbon leakage within the group of the Annex I countries. Although these additional emissions are technically counted as leakages,[2] they have no influence on carbon concentration in the long term, due to the "banking" provision of the Protocol (see page 18).
- Some model scenarios predict a fall of the international oil price relative to the coal price in response to the implementation of the Kyoto Protocol, leading to a shift of energy demand from coal to oil and a corresponding fall of the carbon intensity in some coal consuming countries (inducing "negative leakages" as emissions in some non-Annex I countries – like China – would fall when the Protocol is implemented in Annex I countries). Negative leakages would be most likely to appear if the supply elasticity of oil is small while the supply of coal is elastic.
- Finally, other factors may influence the size of the leakages such as adverse income effects in energy-exporting economies or different degrees of market integration depending on the baseline scenario. These factors are likely to exert a second order influence only over the time horizon of the first budget period of the Protocol.

Some analytical results

As the above discussion highlights, carbon leakages result from various interactions between different markets and are therefore well suited for analysis by a General Equilibrium model. Based on a simple, static CGE model, Burniaux and Oliveira-Martins (1999) have tried to identify which of the parameters mentioned above are the most influential in determining the rate of carbon leakage. The main conclusions are:

- The non-energy trade channel is less influential in determining leakages than often thought. Except for low values of the trade substitution elasticities (below 2) at which leakage becomes very low, the leakage rate is not very sensitive to the degree of substitution on the non-energy markets. Similarly, the degree of international capital mobility does not affect leakages significantly. With a more elaborate description of the international capital markets, the G-Cubed model reports a similar result. It indicates that most of the capital reallocation induced by the implementation of the Kyoto Protocol would take place among Annex I countries rather than towards non-Annex I countries, therefore contributing little to carbon leakage (McKibbin *et al.*, 1999).

Box 4. Carbon leakage: an unresolved issue (*cont.*)

- By far the most influential parameter is the supply elasticity of carbon. Elasticity values above 4-5 yields small and relatively stable leakage rates. But when the carbon supply elasticity falls below 4, leakage rises very substantially (to rates of up to around 40 per cent). As an extreme case, which illustrates the intuition behind this result, a perfectly inelastic supply of carbon would make it impossible to reduce world emissions of carbon and the leakage rate of any unilateral abatement would, by definition, equal 100 per cent. Correspondingly, with an elastic supply of carbon – and coal in particular – over the medium term, as embodied in the GREEN model, carbon leakages are small.
- The supply elasticity of carbon is more influential than the degree of substitution on the international carbon market. In GREEN, coal is treated as an almost homogenous commodity but, as Table 6 illustrates, this assumption does not influence the leakage rate. The table shows that a high supply elasticity of coal always generates a low leakage rate whatever the imputed degree of substitution on the international coal market.[3] The degree of differentiation on the coal market is however more in influential with a low value of the supply elasticity of coal (see the scenario S2 in Table 6).

Table 6. **Leakage rates in the Kyoto Protocol under various assumptions**

	BaU	S1	S2	S3	S4	S5
No permit trading	4.8%	22.9%	12.6%	27.3%	4.6%	5.1%
Permit trading	2.2%	17.4%	8.7%	21.5%	2.1%	2.5%

BaU: Baseline specification with infinite (downward) supply elasticity of coal, supply elasticity of oil = 2, trade substitution elasticities for coal = 4-5 and oil treated as an homogenous commodity.
S1: Supply elasticity of coal set at 0.1 all other parameters being as in BaU.
S2: Supply elasticity of coal at 0.1 and trade substitution elasticity for coal = 0.5, all other parameters as in BaU.
S3: Supply elasticity of coal set at 0.1, supply elasticity of oil set at 0.5; all other parameters as in BaU.
S4: Trade substitution elasticities for coal = 0.5; all other parameters as in BaU.
S5: Trade substitution elasticities for coal = 10; all other parameters as in BaU.
Source: Burniaux and Oliveira Martins (1999).

- Finally, the degree of substitution in the production function also matters for the size of the carbon leakage, a fact that has attracted little attention so far in the analytical literature. Higher factor substitution yields larger carbon leakages as it amplifies the reduction of the carbon demand and the increase of demands for other factors, including carbon-free energy sources, leading to higher adjustments of the international prices. The sensitivity analysis performed with the WorldScan model confirms this result (Bollen *et al.*, 1999).

Box 4. **Carbon leakage: an unresolved issue** (*cont.*)

The results from the GREEN model suggest that carbon leakage is small for the range of parameters most frequently cited in the literature (and used in GREEN). In particular, this result reflects the presumption that the supply of coal is fairly elastic over the medium term. Waves of coalmines closing down in Europe and other OECD countries in recent years suggest that this presumption is justified. However, more empirical evidence about the supply response of coal and oil producers would clearly be needed to make this statement with greater confidence.

1. High transportation costs, lack of infrastructure and other technical aspects have so far contributed to restrict coal trading to a fraction of the world coal production, although Kolstad *et al.*, (1999) argue that the international coal market is more integrated than it looks.
2. Both Bollen *et al.* (1999) and the OECD GREEN calculations treat within-Annex 1 leakages in this way.
3. Thus, the uncertainty about the degree of integration of the international coal market only matters when the supply of carbon is inelastic, as pointed out by Kolstad *et al.* (1999).

cost of meeting global emission targets by nearly 60 per cent. Table 7 suggests that "where" flexibility could reduce the marginal costs of meeting the Kyoto targets by more than half. Average costs could fall substantially too. Gielen and Koopman (1998) estimate that permit trading could reduce the total cost for Annex 1 countries of implementing the Kyoto Protocol from 0.25 per cent of real consumption to 0.10 per cent. Based on simulation results for the United States alone, Manne and Richels (1998) report economic cost falling from more than $80 billions (1 per cent of US GDP) to $50 billions (both at 1990 prices), a reduction of more than 40 per cent.[52] Simulations with the OECD GREEN model suggest that the total GDP loss for OECD countries could be cut by more than a third (slightly less for real income) to become virtually insignificant (close to 0.1 per cent by 2010, virtually the same as for Annex 1 countries as a whole, see Figure 3, panel A). When interpreting this result, it should be borne in mind that it reflects very substantial gains from trading but also that the losses even in the absence of trading are fairly low due, *inter alia*, to substantial terms-of-trade gains among OECD countries associated with lower energy prices and the fall of energy imports.

About 80 per cent of the gains from trading arise from greater efficiency – GHG emission reductions are made where they are less costly, which in practice means mainly in the CIS. The remaining gains come from the fact that aggregate Annex 1 emissions are higher under a permit trading scheme than if countries meet their target individually, with the difference being equal to the gap between the CIS commitment and emission levels in the BaU scenario. With less overall Annex 1 abatement,

Table 7. **Marginal abatement costs without and with emission trading, 2010**

1995 $ per ton of carbon

		Without trade	With trade
United States	WorldScan[1]	41	18
	CETA[12]	191	51
	G-Cubed[2]	82	57
	AIM[4]	170	72
	RICE[10]	148	72
	MS-MRT[7]	272	82
	MIT-EPPA[8]	216	82
	GREEN	**231**	**90**
	SGM[9]	211	93
	GTEM[5]	365	123
	Merge[3]	304	154
	Oxford Model[6]	464	247
Western Europe	WorldScan[1]	83	18
	G-Cubed[2]	253	57
	AIM[4]	226	72
	RICE[10]	184	72
	MS-MRT[7]	200	82
	MIT-EPPA[8]	319	82
	GREEN	**189**	**90**
	SGM[9]	452	93
	GTEM[5]	760	123
	Oxford Model[6]	1 077	247
Japan	WorldScan[1]	93	18
	G-Cubed[2]	106	57
	AIM[4]	266	72
	RICE[10]	279	72
	GRAPE[11]	356	80
	MS-MRT[7]	452	82
	MIT-EPPA[8]	576	82
	GREEN	**182**	**90**
	SGM[9]	411	93
	GTEM[5]	733	123
	Merge[3]	559	154
	Oxford Model[6]	1 213	247

1. Emission reductions and economic costs are from Gielen and Koopmans (1998); carbon taxes are from Bollen, *et al.* (1998).
2. Global General Equilibrium Growth Model, Australian National University, University of Texas and US Environment Protection Agency; derived on the basis of Weyant and Hill (1999), Figures 7, 8, and 9.
3. Model for Evaluating Regional and Global Effects of GHG Reductions Policies, Stanford University and Electric Power Research Institute; derived on the basis of Weyant and Hill (1999), Figures 7, 8, and 9.
4. Asian Pacific Integrated Model, National Institute for Environmental Studies (NIES-Japan) and Kyoto University; derived on the basis of Weyant and Hill (1999), Figures 7, 8, and 9. Figures for the CIS are quoted from Kainuma, Matsuoka and Morita in OECD (1998*a*), 1009, p. 167.
5. Global Trade and Environment Model, Australian Bureau of Agriculture and Resource Economics (ABARE, Australia); derived on the basis of Weyant and Hill (1999), Figures 7, 8, and 9. Figures for the CIS and Eastern Europe are quoted from Tulpulé *et al.* in OECD (1998*a*).
6. Oxford Economic Forecasting model; derived on the basis of Weyant and Hill (1999), Figures 7, 8, and 9.
7. Multi-Sector, Multi-Region Trade Model, Charles River Associates and University of Colorado; derived on the basis of Weyant and Hill (1999), Figures 7, 8, and 9.
8. Emissions Projections and Policy Analysis Model, Massachusetts Institute of Technology; derived on the basis of Weyant and Hill (1999), Figures 7, 8, and 9.
9. Second Generation Model, Batelle Pacific Northwest National Laboratory; derived on the basis of Weyant and Hill (1999), Figures 7, 8, and 9.
10. Regional Integrated Climate and Economy Model, Yale University; derived on the basis of Weyant and Hill (1999), Figures 7, 8, and 9.
11. Global Relationship Assessment to Protect the Environment Model, Institute for Applied Energy (Japan), Research Institute of Innovative Technology for Earth (Japan) and University of Tokyo; derived on the basis of Weyant and Hill (1999), Figures 7, 8, and 9.
12. Carbon Emissions Trajectory Assessment Model, Electric Power Research Institute and Teisberg Associates; derived on the basis of Weyant and Hill (1999), Figures 7, 8 and 9.

Figure 3. **Implementing the Protocol: GDP and real income changes under alternative permit trading regimes, 2010**

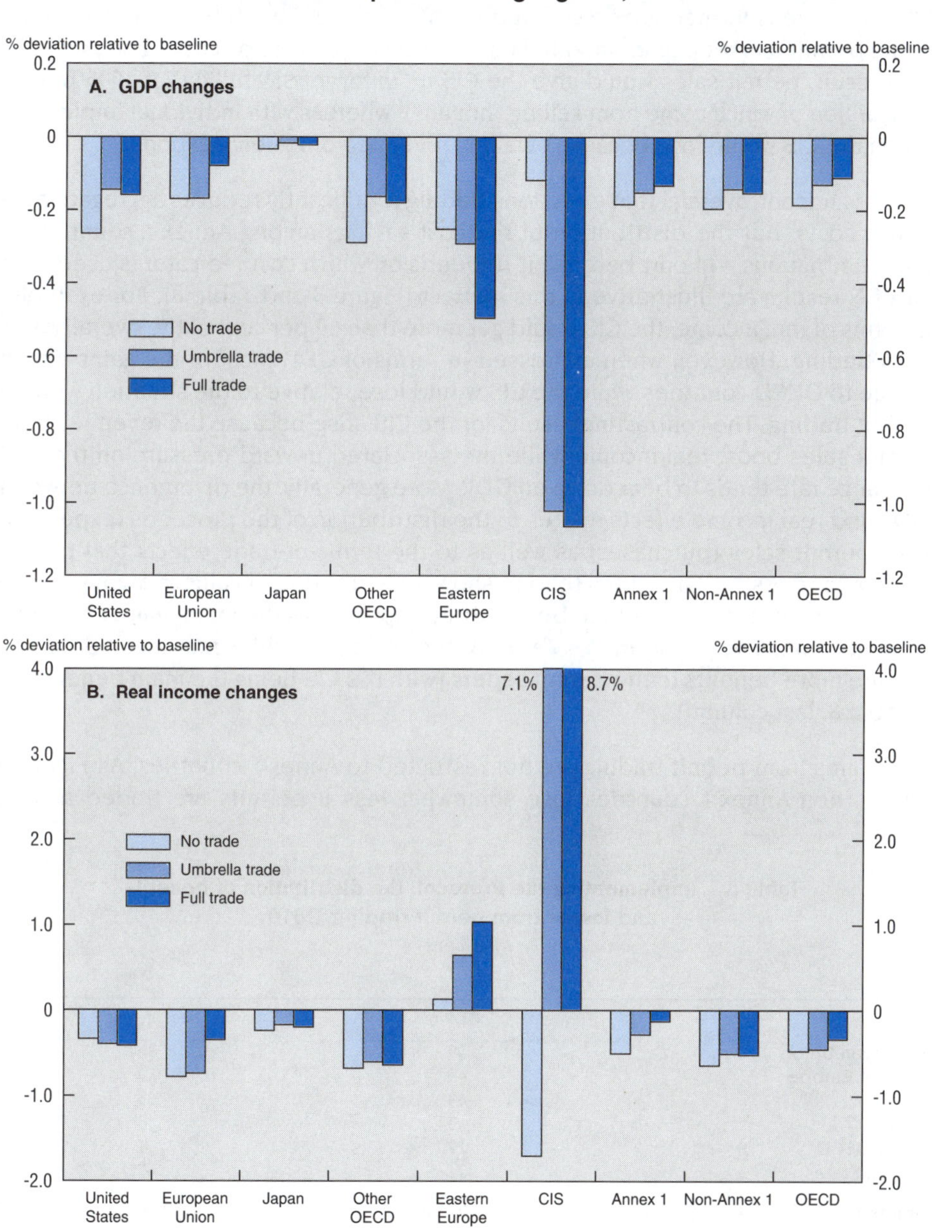

Source: GREEN Model, OECD Secretariat.

the aggregate cost is also lower. Scenarios with the GREEN model suggests that, by 2010, the CIS could sell some 130 million tons of carbon as "hot air" and about 280 million tons further as emission reductions, making the CIS the dominant seller in the permit market (some potential ramifications of this are discussed on page 50). As a result, permit sales would give the CIS an inflow of $39 billion (at 1995 prices) ($12 billion of which come from selling "hot air"), whereas with individual implementation the CIS would lose because of falling revenue from energy exports.

It is uncontroversial that emissions trading significantly reduces aggregate economic costs, but the distribution of the cost savings among Annex I countries is more ambiguous – in part because it depends on which cost indicator is used. The GREEN results are illustrative in this respect (Figure 3 and Table 8). For example, in terms of real income, the CIS would get more than 60 per cent of the overall gains from trading. However, when expressed in terms of GDP, most of the gains would accrue to OECD countries while the CIS would lose relative to the situation with no permit trading. The contrasting results for the CIS arise because the revenues from permit sales boost real income while the associated upward pressure on the real exchange rate tends to bear down on GDP. More generally, the divergence between GDP and real income effects relates to the distribution of the proceeds (expenses) from permit sales (purchases) as well as to the terms-of-trade effects that permit trade generates.[53] GDP figures tend to show a distribution of gains in favour of permit buyers (in terms of GDP, the United States appears as the major gainer because of large purchases of permits – Table 8, first column), while real income figures impute more benefits to the permit sellers (with the CIS being the main beneficiary – Table 8, last column).

Gains from permit trading are not restricted to Annex I countries. As Figure 3 shows, non-Annex I countries lose somewhat less if permits are traded among

Table 8. **Implementing the Protocol: the distribution of benefits and losses from permit trading, 2010**

Per cent

	GDP	Real income
European Union	59	36
Eastern Europe	–9	4
CIS	–43	63
Japan	5	3
Other OECD	17	1
United States	71	–7
Annex I	**100**	**100**

Source: GREEN Model, OECD Secretariat.

Annex I countries than if they meet their commitments individually. The explanation is that permit trading changes the nature of the cut in energy consumption; it shifts part of the burden of this cut from imported oil to CIS coal production. As a result, oil exporting countries are much less affected by the emission reductions in Annex I countries. A related effect is that the leakage rate with tradable permits is likely to be lower than under the individual approach.

- Reasons why "Where" flexibility may be restricted

The unlimited "where" flexibility is effective in lowering costs but also raises a number of questions and difficulties:

- First, there is no consensus among Annex I Parties on how to meet the targets, and some Parties may be politically disinclined to rely on market-based instruments such as permit trading. In the extreme, this could lead some countries or groups of countries to refrain from trading, which again could lead to some degree of market segmentation (see page 46).
- Second, the proportion of trading relative to commitments may cause political concern. The full-trade scenario indicates that OECD countries would rely to a considerable extent on buying permits to satisfy their obligations: for example, the proportion of emission reductions met by purchasing permits could range from 32 per cent in Japan to 42 per cent in the European Union. This may fuel the feeling that OECD countries seek to transform the burden of mitigation into capital transfers to the CIS and may be thought to conflict with Article 17 of the Protocol which specifies that "trading shall be *supplemental* to domestic actions for the purpose of meeting quantified emission limitation and reduction commitments". These concerns could lead to general restrictions on how much individual countries can trade, as exemplified by the proposal of the European Union to put caps on permit purchases[54] (see page 47).
- Third, the high proportion of the CIS Assigned Amount which is sold to other Annex I countries (43 per cent in 2010, according to the GREEN scenarios) may undermine the credibility of the Protocol if it turns out that these emission reductions cannot be achieved or effectively verified. More generally, the measurement and other implementation problems discussed in Chapter 2 are sometimes taken as an argument for restrictions on trade.
- Fourth, the large amount of "hot air" may also create perception problems even if, as described in Chapter 2, the banking provisions of the Protocol imply that this may have little impact on cumulative emissions over a longer time period.

– Fifth, the fact that the CIS taken as a whole would be the sole seller of allowances raises the possibility of monopolistic behaviour and associated efficiency loss in the market for permits (see page 50).

– Finally, developing countries typically refuse to consider quantitative emission limits which restrict their *per capita* emission rates to levels far below those of developed countries unless the developed countries are willing to show some commitment to reduce their own emissions. Taking this position as given, it may make sense to insist that Annex I countries do undertake "unnecessarily" high abatement costs in order to induce some major developing countries to accept quantitative limits (see further in Chapter 4).

Against this background, the following three sub-sections review various possible departures from the full "where" flexibility and their impact on economic costs.

• Trade segmentation

Trade segmentation refers to the possibility that some Annex I Parties do not participate in general permit trading even if they may trade among themselves in a separate trading area (a so-called "double-bubble" scenario). Under this kind of scenario, there would be divergences between marginal abatement costs in the different segments of the permits market. Parties who would be net buyers of permits under full trading, but who do not wish to trade permits, have to abate more domestically and at a higher cost. Other net buyers would gain as the price of permits would be lower than with full trade.

The "double-bubble" scenario has been simulated using various models, for instance, by assuming that the European Union and Eastern European countries would not participate in permit trading.[55] This particular scenario may not be very realistic but it illustrates the mechanisms at work. The results point to substantially lower efficiency gains with segmented trade than with full permit trading among Annex I countries. According to the GREEN model ("umbrella" scenario in Figure 3), restricting permit trading to the so-called "umbrella countries" (North America, the Pacific countries and the CIS) would reduce overall real income gains from trading by 40 per cent (relative to the full Annex I trading scenario). The real income losses are concentrated in the European Union, which forgoes the benefit of buying permits, in the CIS, which sells fewer permits at a lower price, as well as in Eastern Europe. On the other hand, countries of the "umbrella" group gain slightly more than in the full trade scenario since the permit price is lower in the absence of the European Union as a large buyer on the market. Thus, by not participating in trading, Parties would tend to suffer higher costs while conferring additional benefits on Parties which trade.

- Trade restrictions

The condition in the Kyoto Protocol that emissions trading should be supplemental to domestic action raises the possibility that restrictions be imposed on the amount of emissions Parties can buy or sell. The recent proposal by the European Union (European Union – The Council, 1999) provides an example of how supplementarity is interpreted in practice. The proposal puts ceilings on the amounts of emissions that Annex 1 countries could exchange under the three flexibility mechanisms. Thus, the ceilings apply to permit purchases and "project-based" net acquisitions of emissions as well as permit sales and "project-based" net transfers of emissions. It is very difficult at this stage to provide any quantified assessment of the impact of these proposed restrictions, including on the size of the efficiency gains from using the flexibility mechanism.[56] As a more modest contribution, the following analysis aims to illustrate the impact of restrictions by reviewing some recent results in the literature and presenting some counterfactual scenarios of ceilings on emission purchases and sales.

Bollen *et al.* (1998) have simulated restrictions on permit purchases on the WorldScan model. Under the assumption that purchases of permits by Annex 1 Parties cannot exceed 10 per cent of their QELRCs, all OECD countries would be rationed buyers and their marginal costs would therefore differ (although to a lesser extent than with no trade). Manne and Richels (1998) have also simulated the impact of limitations on permit purchases and show that not allowing buyers to satisfy more than one third of their reduction commitment with allowances purchased from abroad could more than double the GDP loss incurred by the United States.[57] According to the GREEN simulations presented above, unrestricted emission trading among Annex 1 countries would imply substantial purchases of emission rights by OECD countries from the CIS and Eastern European Countries. By 2010, these purchases of emissions would range from 30 to more than 40 per cent of OECD country reductions relative to baseline levels. Figure 4 reports results from two alternative scenarios where buying countries were limited to meet, respectively, 30 per cent and 15 per cent of their reduction requirements through permit purchases.[58] The 30 per cent restriction has a rather small impact on the cost-saving from trading as it only marginally reduces permit purchases.[59] In contrast, the 15 per cent ceiling would cut the expected gains from trading for the whole Annex 1 area by half.[60]

A restriction on permit sales by Annex 1 Parties would have broadly parallel overall effects. It would limit abatement in low-cost countries and, by reducing the amount of permits to be sold abroad, force high-cost countries to reduce further their domestic emissions. Marginal abatement costs among Annex 1 countries would diverge and overall efficiency be reduced. With unrestricted trade among Annex 1 countries, GREEN results suggest that the CIS would sell 43 per cent of its

Figure 4. **The impact of restrictions on permit purchases**

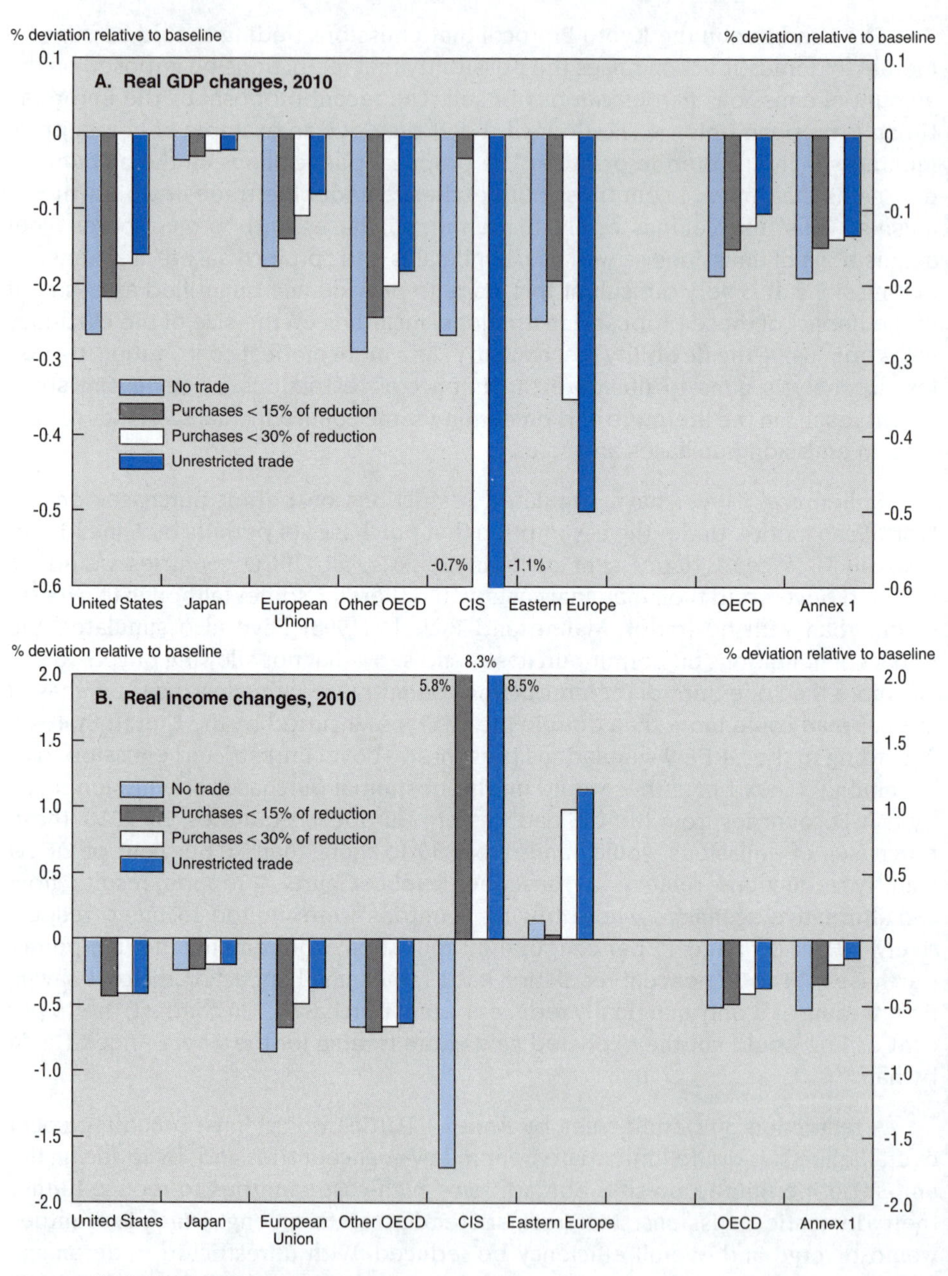

Source: GREEN Model, OECD Secretariat.

Figure 5. **The impact of restrictions on permit sales**

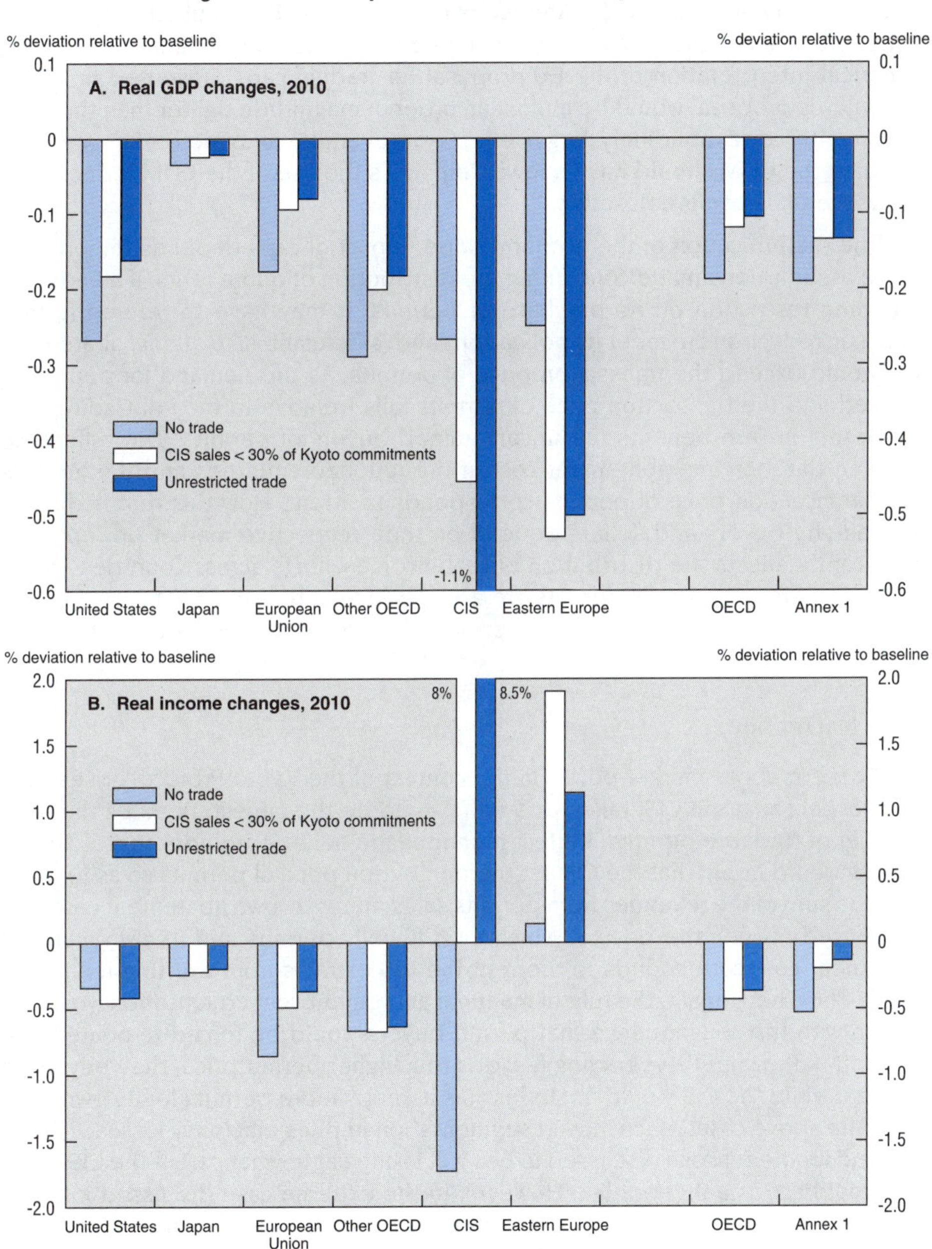

Source: GREEN Model, OECD Secretariat.

total Kyoto commitment by 2010. Limiting this proportion to 30 per cent would have only a marginal impact for Annex 1 countries overall but would imply a reduction of the real income gain of OECD countries by 40 per cent (Figure 5).[61] On the numerical interpretation of the EU proposal for trading caps presented in OECD (1999*b*), this proposal would be almost an order of magnitude tighter than this simulation with correspondingly larger effects on overall economic costs.[62] Ceilings including the CDM should have a lower impact as the use of the CDM reduces the role of the CIS as emission seller.

The quantification of the economy-wide impact of caps depends importantly on the assumptions made concerning the distribution of quota rents. If a Party has a binding restriction on its purchase of permits, it may have to rely on further domestic reductions to meet its obligations and, as a result, its marginal abatement cost would exceed the transaction price of permits. As the demand for permits is reduced and the transaction price of permits falls (relative to the full trading scenario), this in turn benefits the unconstrained buyers of permits. The difference between the marginal abatement cost in the rationed countries and the international transaction price of permits corresponds to a rent. How this rent is distributed among buyers and sellers depend on their respective market power,[63] and significantly affects the distribution of real income effects across countries. In the above scenarios, sellers of permits are assumed to capture the quota rents based on the likelihood that the seller side may show greater concentration than the buyer side.

- Market power

Concern about market power in the context of the Kyoto Protocol arises from the prospect that the CIS, taken as a unit, would be the largest, perhaps the sole, supplier of tradable permits. Perfect monopolistic behaviour under these circumstances would imply that the CIS set the transaction price of permits so as to maximise the sum of the revenues from permits sales minus its own abatement cost. The difference between the price at which the CIS sells permits and its own marginal abatement cost corresponds to the rent the CIS extracts from exerting its market power. Thus, here again, the rule of marginal abatement cost equalisation would be broken with the consequence that permit buyers would be forced to abate more than with full competitive trading (as, due to a higher permit price, they buy fewer permits) while the CIS would abate less (as it keeps more permits for its own use). As for the above cases, such market segmentation implies efficiency losses. In practice, and for the reasons discussed in Box 5, it is unclear to what extent the CIS could and would act as a monopolist. However, in the extreme case discussed in Box 5 where the CIS is able to exert full market power, the permit price in 2010 could be some 20 per cent higher than under perfect competition and OECD countries' gain from trading permits could be reduced by a fifth or more (Burniaux, 1998).

Box 5. How important is market power in the market for emission permits?

The simulation results on the effects of permit trading reported in the main text are predicated on the assumption of perfect competition in the market for permits. This assumption is not necessarily realistic. Concern about imperfect competition in permit trading arises from a result which has been confirmed by virtually all model quantifications of the Kyoto protocol: permit trading under the conditions set up in Kyoto is likely to be overwhelmingly dominated by the Russian Federation and Ukraine (referred to as the CIS). Figure 6 shows the trade patterns obtained with the OECD GREEN model: the CIS would sell about 450 million tons of carbon by the first budget period and thereby be the far dominant seller. An important part of these sales would be accounted for by "hot air". Simulations extending the Kyoto commitments over a longer period show that the CIS would remain the major supplier of permits up to 2030, when the United States would emerge as another major permit seller. Thus, conditions exist for the CIS to behave monopolistically. This Box provides benchmark values for the impacts on trade and abatement costs under the extreme assumption that the CIS would be able to exploit fully its market power in permit trading.

A model of monopolistic competition in permits trading

The approach used here extends the model formulated by Hahn (1984) at the firm level to Annex 1 countries. It assumes that one dominant country has the ability to influence the price of permits. In the case of monopoly, this country sets the permit price at the level which corresponds to the maximisation of the difference between revenues from permit sales and its abatement costs. All other countries behave as price takers, *i.e.* they minimise their abatement/trading costs given the permit price set by the dominant country. There is one single transaction price for permits (thus, permits are traded as if they were a homogenous good). The country with market power is identified as the one which sells or purchases the largest share of permits in the corresponding competitive scenario (where all countries are price takers).

The first-order conditions of the optimisation problem of the monopolistic country (country 1 below, leads to the following specification for the permit transaction price (P) (see Burniaux (1999) for further details):

$$[1] \quad P = \frac{\left(L - \sum_{i=2}^{m} E_i(P) - E_1^0 \right)}{\sum_{i=2}^{m} E_i'(P)} - C_1'(E_1)$$

Box 5. How important is market power in the market for emission permits? (*cont.*)

Figure 6. Permit trading by country/region under the Kyoto Protocol, 2010

A. Permit purchases by country/region

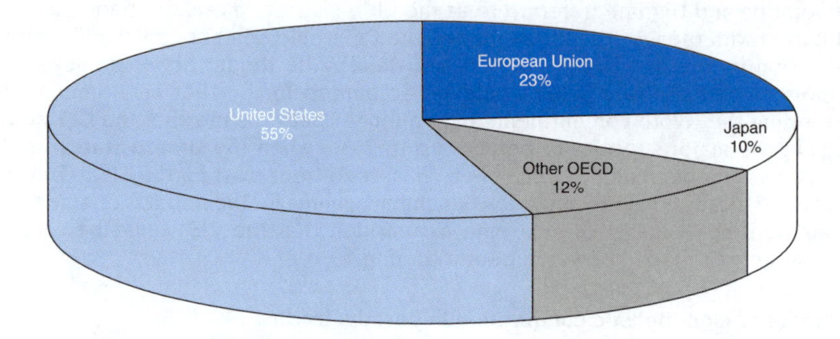

B. Permit sales by origin

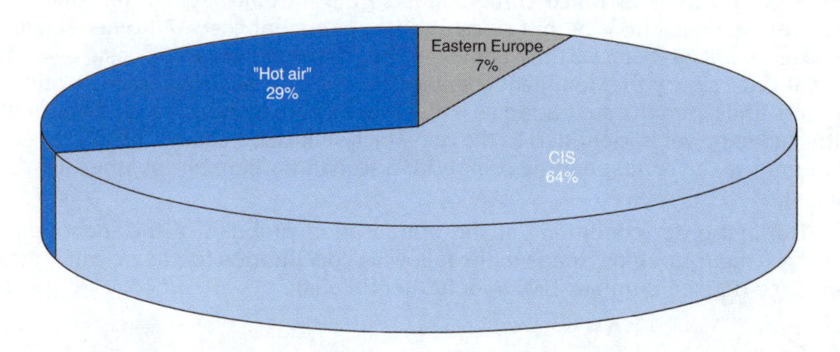

Source: GREEN Model, OECD Secretariat.

Box 5. How important is market power in the market for emission permits? (*cont.*)

In equation [1], the price set by the monopolistic (or monopsonistic) country (P) is calculated by adding a positive (or negative) mark-up to the marginal abatement cost of the country - $C_1'(E_1)$. This mark-up, in turn, is a function of:

1. The excess demand for permits by country 1 relative to its initial permit allocation E_1^0 (as indicated by the numerator of the right-hand fraction with the emissions of country 1 being calculated as a residual between the total permits quota L and the total emissions from price-taking countries 2 ....m). In the case of monopoly, this excess demand is negative.
2. The inverse of the sum of first derivatives of demands for permits in the price-taking countries 2 to m (with $E_i'(P)$ being the first derivative of the demand for permits (or emissions) as a function of the permit price P in country i).

Equation [1] has the following implications:

1. If the price-setting country uses less permits than its initial allocation, then: $\left(L-\sum_{i=2}^{m} E_i(P)-E_1^0\right)<0$. The country sells excess permits at a price set above its marginal abatement cost (both $C_1'(E_1)$ and $E_i'(P)$ are negative). Compared with the competitive case, the monopolist sells fewer permits (but at a higher price) and, thus, abates too little.
2. If the price-setting country uses more permits than its initial permit allocation, then $\left(L-\sum_{i=2}^{m} E_i(P)-E_1^0\right)>0$. The country buys permits at a price set below its marginal abatement costs. Relative to the competitive case, the monopsonist buys less permits (but at a lower price) and, thus, abates too much.
3. It follows from (2) and (3) that the total abatement cost is always larger than under the competitive scenario.
4. The price set by the dominant country – and the resulting loss of efficiency relative to the competitive case – depends on the amount of permits initially allocated to this country (E_1^0 in the numerator of the right-hand ratio). Thus, imperfect competition establishes a link between distributional aspects and overall efficiency. In other words, one initial allocation of emission rights may prove more or less costly than another. This contrasts with the competitive case where the total abatement cost is the same whatever the initial allocation of emission rights.
5. The deviation between the transaction price set by the dominant country and its marginal abatement cost depends on the price elasticity of demand for permits. Thus, the easier it becomes to substitute carbon-free energy sources for carbon-based ones – for instance when backstop options become available – the more difficult it becomes for the price-setting country to exert market power.

Box 5. How important is market power in the market for emission permits? (*cont.*)

Main quantitative results

Simulations based on the GREEN model, in which the CIS is assumed to exert its market power indicate that, by 2010, the permit price set by the CIS is 21 per cent higher than the corresponding competitive price, (91 1995 $ per ton of carbon compared with 75 1995 $). At this higher price, the CIS would sell less permits and therefore abate less than in the competitive scenario. As a result, the marginal abatement cost in the CIS would amount in 2010 to less than 30 1995 $ per ton of carbon, much below both the marginal cost with the perfect competition and the price at which permits are sold with imperfect competition. The wedge between the transaction price set by the CIS and its marginal abatement cost is the mark-up on permit sales; it reflects the market power of the CIS in permit trading. It also corresponds to the difference between marginal abatement costs in the CIS and in other Annex 1 countries (as other Annex 1 countries act as price takers and equalise their marginal costs with the permit price set by the CIS). In 2010, this mark-up reaches a rate of 178 per cent (*i.e.* the permit price set by the CIS is almost three times its marginal abatement cost) and declines steadily thereafter (under the assumption the Kyoto targets remain unchanged).

Imperfect competition implies that the total emission reduction is achieved at a higher cost than in the perfectly competitive case. This results from the fact that the CIS (the low cost country) abates too little while the other Annex 1 countries (the high cost countries) abate too much compared with the competitive scenario. The total abatement cost in OECD countries is higher than in the competitive trade scenario, albeit much lower than in the no-trade situation (Figure 7). Price manipulation by the CIS would thus reduce the cost saving expected from permit trading. In 2010, competitive permit trading would cut OECD costs by half, but due to monopolistic price setting, this cost gain is reduced by a fifth.

Summing-up

The above results suggest that monopolistic trading of permits may reduce the total efficiency gains from trade by about a fifth. Nevertheless, there is no possibility that monopolistic behaviour completely offsets the efficiency gains from trade. This analysis relies on three important – and possibly questionable – assumptions about the market structure for permits: first, that permit trading in the CIS is centralised rather than being pursued by individual countries and firms; second, that trade in other Annex 1 countries is performed by individual firms which have no market power; and, third, that the CIS does not take into account the effects of its monopolistic behaviour in the permit market on energy prices and thereby its terms of trade.* In addition, it is assumed that firms have no market power on the goods market. It may seem unlikely that permit trading under the Kyoto protocol will fully correspond to this particular structure. The results might therefore best be seen as providing an upper bound estimate of the inefficiency which could be generated by price manipulation if the CIS were to exert monopolistic power in permits trading.

Box 5. How important is market power in the market for emission permits? (*cont.*)

Figure 7. **Abatement cost[1] of OECD countries[2] under alternative market structures, 2010**

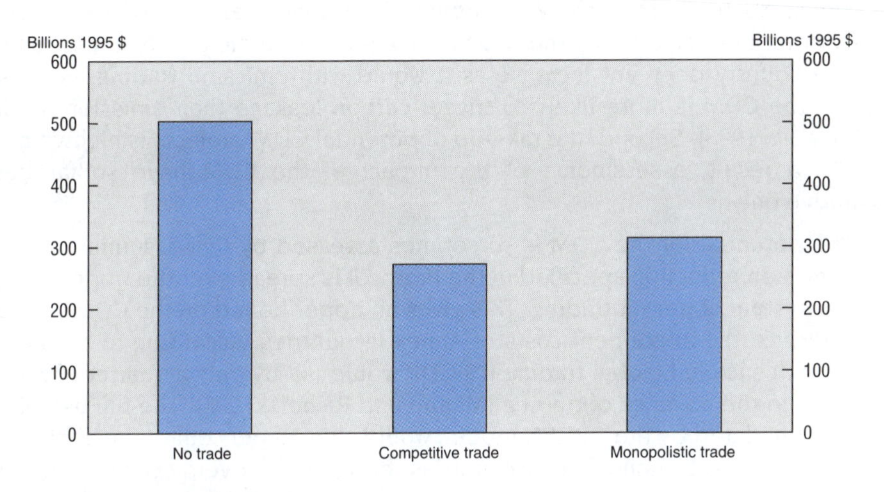

1. Calculated as the marginal abatement cost times the emission volume.
2. Including Eastern European countries.

Source: Green model, OECD Secretariat.

* In the simulations, overall emission cuts are given. However, the price of permits will affect the fuel mix in individual countries and thereby the relative prices of different types of energy. This again may affect the terms of trade of the CIS which exports mainly oil and gas.

- "Where" flexibility with the rest of the world.

The Clean Development Mechanism (CDM) allows Annex I countries to count emission reductions from projects implemented in non-Annex I countries as part of their obligation in the first commitment period (2008-12) as long as these reductions are "certified" (see discussion in Chapter 2). In principle, CDM investments rely on the same economic incentives as emission trading: investors will undertake projects in which the abatement cost is lower than in other locations – including their own country. In that sense, the allocation of emission reductions obtained with CDM is cost-effective. Quantifying the effects of CDM is difficult, however. First, CDM projects cannot be modelled in the same way as emission trading because CDM, unlike emission trading, is not subject to any kind of quota on host country emissions. Abatement is achieved through the replacement of existing capacity mostly or entirely paid for by the Annex 1 investor, while the cost of energy use in the host country does not increase as it would with emission trading. In consequence, the CDM is more likely to trigger carbon leakage than emission trading (Bollen *et al.*, 1998). Second, the take-up of potential CDM projects is highly uncertain. As a result, assessments of the impact of the CDM have so far been illustrative only.

The potential for the CDM is sometimes assessed by first assuming that the total emission reduction specified in the Protocol is spread over the world through a global system of permit trading. This gives an upper bound on the CDM's potential to reduce the abatement costs of Annex 1 countries. According to the Merge model, with such full global trading US GDP would fall by only a quarter of what it would fall in the absence of trading (Manne and Richels, 1998). The full use of the CDM, as simulated by the GREEN model, would shift 80 per cent of the abatement decided in Kyoto to non-Annex 1 countries, bringing the overall costs to Annex 1 countries to very close to zero and the permit price to 10 dollars at 1995 prices per ton of carbon (as compared with average marginal abatement costs of some 200 dollars with no flexibility mechanisms and 90 dollars under Emission Trading and Joint Implementation only). But the practical difficulties in implementing the CDM and in certifying the emission reductions from CDM projects make it unlikely that this potential, although very large, could be anywhere near fully exploited. Existing studies approximate these limitations by imposing arbitrary ceilings on the amount of emissions which can be acquired through CDM projects (10 per cent in Bollen *et al.*, 1998; 15 per cent in Manne and Richels, 1998; 20 per cent in US Administration, 1998). Results from Bollen *et al.* (1998) and US Administration (1998) suggest that if the CDM is additional to full emission trading among Annex 1 countries,[64] the cost reduction obtained through the CDM with these restrictions is rather marginal. According to these scenarios, most projects undertaken under the CDM facility would take place in China.

"When" flexibility

The time frame over which the emission reductions decided in Kyoto will be achieved is another source of uncertainty in assessing the economic costs. By setting the first commitment period some time into the future, the Protocol aimed to give Parties enough time to adopt gradual and credible paths for emission reductions. The fact that the targets are stated in terms of average emissions over a five-year period (2008-2012) confers an additional degree of flexibility, and the "banking" provision implies that over-fulfilment of the target will count against future emission targets. Against these measures in favour of credibility and commitment counts the fact that many aspects of the Protocol are not yet fixed, which together with the delay in ratifying the Protocol may encourage Parties to wait before making any serious adjustments. Details on the implementation of the flexibility mechanisms will now not be finally known until COP6 which may take place at the end of 2000. The Protocol itself will come into force only after a significant majority of signatories have ratified it[65] and the ratification process might be somewhat prolonged for some countries.[66] The earliest possible entry into force is thus 2001; even then some procedural hurdles will remain, such as setting up the necessary accounting and verification institutions.

The "when" flexibility in the Protocol implies some degree of freedom to pick a cost-effective timepath towards the Kyoto targets. First, abatement costs may fall over time as new sources of energy with lower or no carbon content become available at low cost. This argument calls for delaying abatement to the later years covered by the Protocol. On the other hand, delaying action increases the magnitude of the effort required to bring emissions down to the levels specified in the Protocol. Moreover, spreading the reductions over a longer time period would reduce costs by reducing the extent of "premature" retirement of existing capacity (though delayed, but credible and predictable adjustment may also reduce this problem). As well, to the extent technological development is endogenous, early abatement may reduce the cost of subsequent efforts. Finally, a gradual phasing in of the reductions may reinforce the credibility of future abatements.

Only some of these aspects are reflected in the GREEN model.[67] Nevertheless, for illustrative purposes the model has been used to simulate three alternative timepaths for different regions individually to meet their Kyoto targets: *i)* a scenario in which emissions reductions are linearly spread over the period from 2000 to 2010; *ii)* a scenario in which the Kyoto targets are achieved by early action spread over the period 2000 to 2005; and, *iii)* a scenario in which late action during the period 2005 to 2010 brings emissions down to their commitment levels (Figure 8). For all regions, spreading the effort over a longer period of time is the least costly option. Reductions phased-in over shorter periods of time clearly generate additional costs associated with accelerated replacement of existing capacity. When considering the time horizon of the Protocol, early and rapid action appears to be the most costly option. This reflects that not all the costs of late and rapid action appear within the Kyoto horizon. Looking instead at costs up to 2050, on the assumption that the Kyoto limitations stay constant, delayed action spread over a five-year period unambiguously appears as the most costly option. Some countries, such as the United States, need very high carbon taxes to meet their commitments

Figure 8. **Implementing the Protocol: costs of alternative timepaths**
(average annual discounted[1] real income losses, per cent)

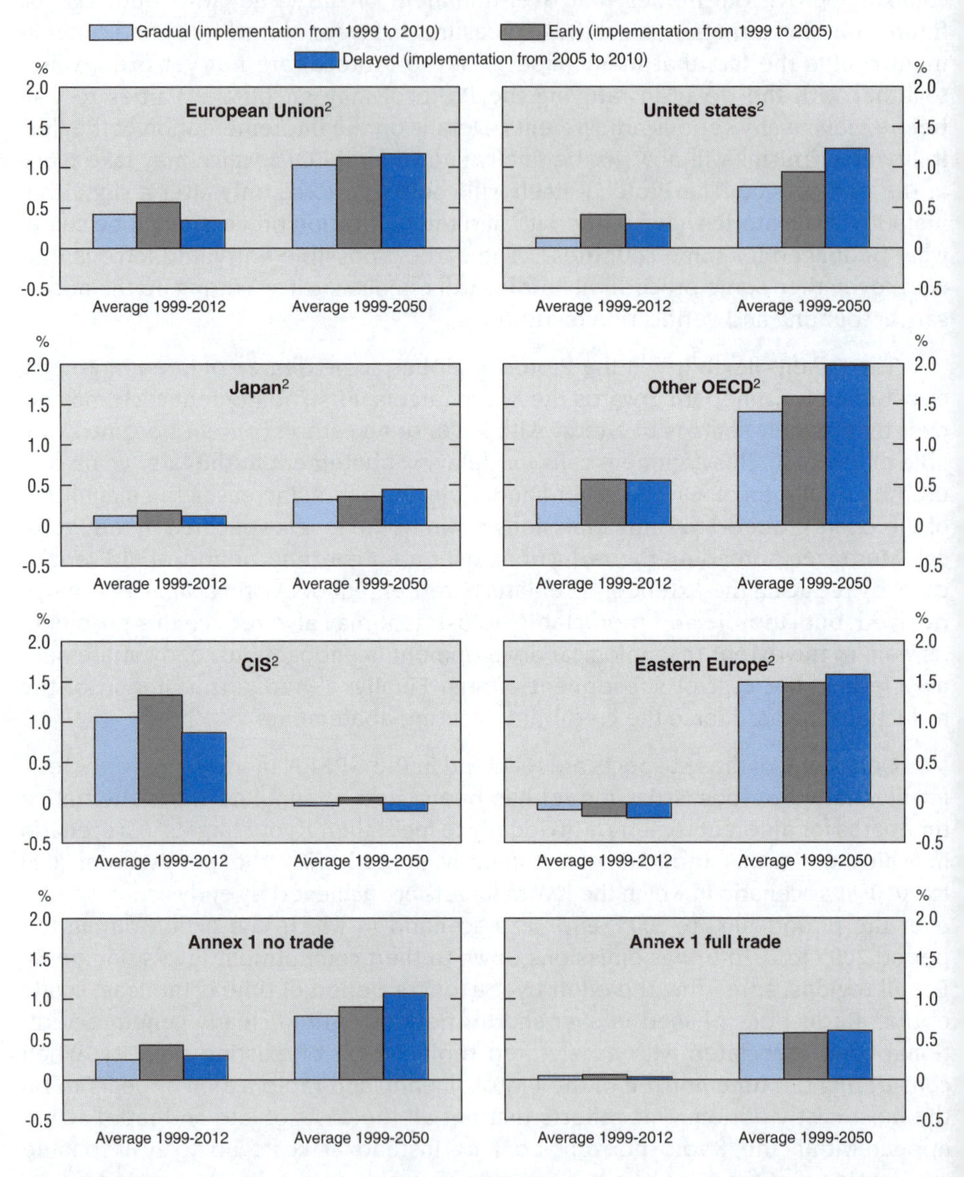

1. The discount rate is 3 per cent.
2. Assuming no permit trade.

Source: GREEN Model, OECD Secretariat.

in time, suggesting that this latter scenario may prove infeasible in practice, given the available technological options.[68] Thus, these results suggest that, within the context of the Protocol, gradually phasing in the reductions from an early date will minimise aggregate costs over time. A factor which crucially affects these results is the level of intertemporal preference.[69] Even more important are the limitations of the model in terms of ignoring forward-looking expectations and endogenous determination of technological change. Against that background, the results should be seen as illustrative only.

"How" flexibility

The "how" flexibility refers to the freedom of Parties to choose among alternative instruments in order to meet their commitments. However, the Protocol specifies that Parties, while achieving their quantified emission limitation, should implement and/or further elaborate policies to progressively reduce or phase-out market imperfections that run counter to the objective of the Protocol (Article 2, v). In particular, this involves the removal of subsidies to energy consumption. Some of the transition economies among Annex 1 Parties have traditionally applied very high subsidies for fossil-fuel consumption. Despite recent reforms, these subsidies remain important in some cases. For instance, a study by the World Bank (1997) points to subsidy rates around 30 per cent in 1995 in the CIS and Eastern European countries.

Removing energy subsidies is a "no regret" measure in the sense that it brings economic benefits beyond the environmental achievement of reduced GHG emissions (though it has distributional consequences including associated adjustment costs). Price distortions generate efficiency losses as producer and consumer decisions are made on the basis of prices which do not represent the economic opportunity cost of resources. It follows that the cost of implementing the Kyoto Protocol would be reduced by eliminating the remaining energy subsidies in the CIS and Eastern European countries. However, the gains would not be restricted to countries in which subsidies are removed. First, removing energy subsidies in the CIS, by reducing energy consumption, would increase the amount of permits that the CIS could sell to other Annex 1 countries if emission rights were tradable. Second, accelerated subsidy reforms in transition economies would reduce international energy prices and hence improve Annex 1 countries' terms of trade and thereby partly offset their cost of meeting the Kyoto targets.

Simulations with the GREEN model confirm that the removal of fossil-fuel subsidies lowers the aggregate economic costs for Annex 1 countries as a group (Figure 9 and Table 9). The resulting efficiency gain is, however, much less substantial that the one which would arise from making emission rights fully tradable. Both with and without permit trading, subsidy reform in the CIS and Eastern Europe may

Figure 9. **Implementing the Protocol: the effect of energy subsidies for costs, 2010**
(real income changes relative to baseline)

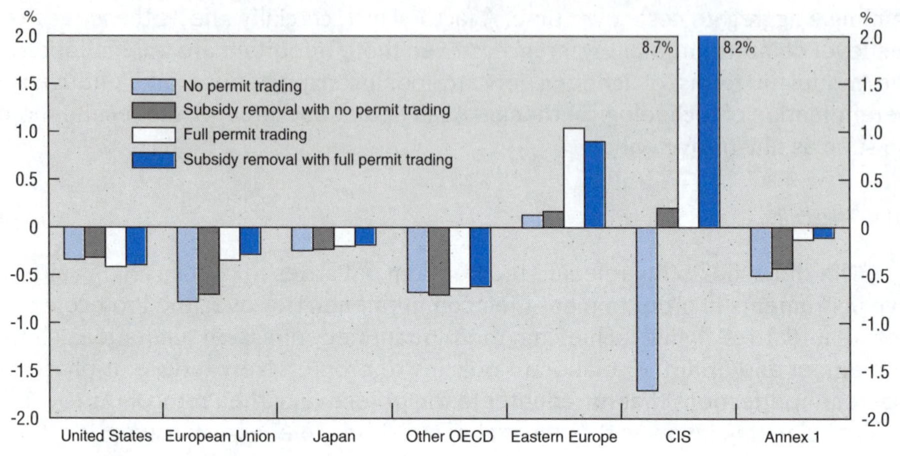

Source: GREEN Model, OECD Secretariat.

Table 9. **The distribution of real income gains from energy subsidy removal, 2010**
Per cent

	No permit trading	Full permit trading
United States	–8.5	34.2
European Union	41.5	108.3
Japan	3.3	16.2
Other OECD	2.2	8.6
Eastern Europe	3.8	–12.1
CIS	57.7	–55.2
Annex 1	100.0	100.0

Source: GREEN Model, OECD Secretariat.

lower the aggregate real income loss of Annex 1 countries by about 15 per cent (from 27 to 23 billion 1995 dollars in the former case and from 107 to 91 billion 1995 dollars in the latter case).

The distribution of the gains from energy market reforms in the CIS and Eastern Europe differs strongly with and without tradable permits. With no permit trading, the CIS is the largest gainer (accounting for 58 per cent of the gains) but there are still substantial positive spillover effects to the OECD countries accruing, not least, as a result of lower energy prices (the United States, Japan and the European Union together receive 44 per cent of the gains: see Table 9). With permit trading, and assuming that this trade has a competitive structure, the efficiency gains from removing subsidies in the CIS are entirely "exported" to the other Annex I Parties as the CIS sells more permits (484 million tons of carbon instead of 440 million tons with no subsidy removal). These permits do not, however, provide any benefit to the CIS. As the aggregate demand for permits remains unchanged, additional supply lowers the permit price so that the revenues the CIS gets from permit sales are slightly lower than in the scenario with no subsidy removal.

Thus, these results suggest that the removal of remaining energy subsidies in transition economies may lower further the costs of emissions reductions. The efficiency gains would not only benefit the countries which undertake the reforms. On the contrary, they would be spread over all OECD countries, in particular the United States and the European Union. Indeed, on some assumptions, the benefit to the countries undertaking reform may be so low as to raise questions about the incentives to do so.

"What" flexibility

"What" flexibility relates to the choice of which emissions to cut and whether to make use of "sinks". As discussed in Chapter 2, the Protocol covers six different types of GHGs, and Parties are free to substitute reductions in emissions of one gas equivalent to increases in emissions in another.[70] There may be large differences of marginal abatement costs across gases – implying significant potential efficiency gains from equalising these marginal costs by substituting among gases. In addition, the Protocol includes net changes of emissions from forestry activities resulting from actions taken since 1990. Here again, Parties may choose reforestation options if they prove less costly than reducing emissions from *e.g.* industrial activities.

The evidence provided in the literature about the cost saving potential of these two options is fragmentary so far. As far as the OECD Secretariat is aware, there are currently very few models that deal with the other GHGs in a comprehensive way.[71] Gielen and Kram (1998) estimate that other gases currently account for 21 per cent of total CO_2 equivalent emitted in Europe and this share is projected to decline up to 2010. According to the same source, the autonomous decline of these gases over time might reduce by 25 per cent the corresponding cut of CO_2 emissions needed to meet the European Union target in 2010. Gielen also suggests that further abatement of non-CO_2 gases can be achieved at relatively low cost (25 ECU per ton of CO_2 equivalent). Results from GTEM (Brown *et al.*, 1999) suggest that the inclusion of methane (CH_4) and nitrous oxide (N_2O) would reduce the marginal abatement cost of meeting the Kyoto targets by a third. To the extent these results prove robust, the costs of meeting the Kyoto targets could be significantly lower than suggested in this paper. However, measurement uncertainties for non-CO_2 gases will have to be reduced considerably; while they remain high, trading may be complicated, particularly as regards sales by countries where non-CO_2 gases are a significant part of total emissions.

The sequestration potential and the economic effects of carbon sinks have not yet been systematically evaluated. Carbon emissions from land-use changes currently amount to 1.6 gigatonnes of carbon per year, compared to 6 gigatonnes of carbon from the burning of fossil fuel. Slowing the rate of loss and degradation of existing forests could reduce CO_2 emissions substantially but most of this potential concerns the tropics rather than Annex 1 countries. On the other hand, there is insufficient time for afforestation/reforestation to significantly change the amount of carbon sequestered above-ground before the first commitment period. Indeed, work by Nilsson and Schopfhauser (1995) indicates that the impact of mass plantation would be significant only after 40 to 50 years. In summary, the sequestration potential of enhancing carbon sinks is likely to be negligible within the current time limits of the Kyoto Protocol, but this potential may become more important if the Protocol is extended in time and country coverage.

4. Beyond the first commitment period: extending Kyoto in time and space

This chapter reviews the scope and policy requirements for moving towards a stabilisation of atmospheric GHG concentrations. Following an analysis which demonstrates the need for a global effort in order to reach this aim, the chapter reviews how burdens could be shared between countries so as to establish international agreement.

4.1. *Long-term aspects of the Kyoto Protocol*

The previous chapter focused on various aspects of the Kyoto Protocol up to the first commitment period of 2008 to 2012. Very few studies provide any quantified assessment of the Protocol beyond this period. Such assessment is very speculative at this stage as it has to be based on assumptions concerning the longer-term targets on which the current version of the Protocol contains no guidance. Despite this uncertainty, it is important that the Protocol be analysed in the context of the longer-term goal of the Framework Convention which is to stabilise the concentrations of GHGs in the earth's atmosphere. This section outlines the implications of an extension into the future of the Kyoto Protocol: *i.e.* the targets decided in Kyoto for Annex 1 countries are maintained unchanged over the longer term, while no specific effort to mitigate emissions is undertaken in non-Annex 1 countries. The outcome of this scenario is that, in the longer term, the costs of maintaining the commitments decided in Kyoto are likely to grow while the potential for emissions trading to reduce these costs would diminish. At the same time, and based on a tentative analysis of the links between CO_2 emissions and atmospheric concentrations, it seems that stabilisation of Annex I emissions at the levels decided in Kyoto is far from sufficient to stabilise concentrations. This forms the background for the subsequent sections on issues involved in achieving a global agreement with the purpose of stabilising concentration levels. The analysis in both this and subsequent sections is couched exclusively in terms of CO_2 given the limited information on other GHGs.

Long-term costs of the Kyoto targets

Figure 10 reports the real income losses relative to the BaU scenario in 2050 under the assumption that Annex I countries individually keep their emissions from exceeding the targets set at Kyoto. The aggregate real income loss for the whole Annex I area reaches 1½ per cent annually (compared with ½ per cent in 2010). Although economic costs increase in all countries, Eastern Europe and the CIS are most affected. The shape of the BaU scenario including the existence of new sources of energy contribute to increase the costs. By 2050, shale oil with a high carbon content is projected to become cheaper than crude oil in the BaU scenario. Applied in this context, a carbon tax – or, equally, a system of tradable permits – would tax shale oil (which has a higher carbon content) more than crude oil. The shift in demand towards oil and the corresponding rise in the international oil price lead to a deterioration in Annex I countries' terms of trade. Conversely, energy exporters would gain, in contrast with their situation in 2010.

In contrast to the first commitment period, emission trading among Annex I Parties would not significantly reduce aggregate costs in 2050. There are two explanations for this, both related to the assumption that, by this time, alternative energy sources (referred to as "backstop technologies") would be available at some uniform price in all countries. First, by harmonising the marginal abatement costs among countries, backstop energy sources reduce the scope for efficiency gains from trading emissions. Second, benefits from trade disappear as increasing returns in abatement occur in high-cost countries (in which backstop energies are competitive) while low-cost countries still use traditional energy sources characterised by decreasing returns to abatement (see Annex 1).

Impact on GHG concentration

The UN Framework Convention on Climate Change (UNFCCC) calls for "stabilisation of greenhouse gas concentrations in the atmosphere at a level that would prevent dangerous anthropogenic interference with the climate system". The Convention however does not specify the concentration values which would correspond to this threshold. Nor has the scientific community determined such a level. Nevertheless, the threshold of 550 ppmv of CO_2 has often been considered as a

Figure 10. **Long-term costs of maintaining the Kyoto targets**
(real income changes in % relative to the baseline)

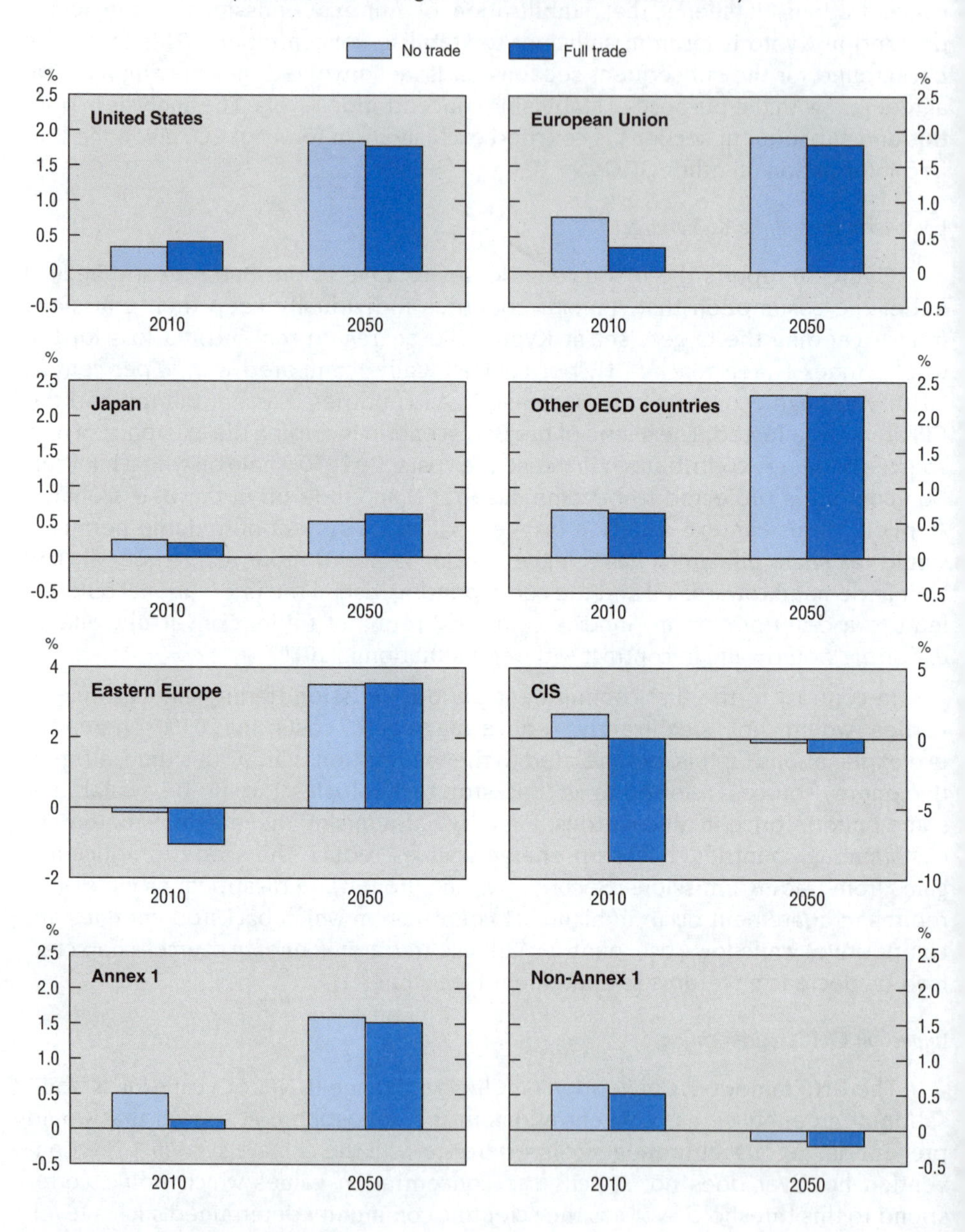

Source: GREEN Model, OECD Secretariat.

target for stabilisation by the end of the 21st century. This level corresponds roughly to twice the CO_2 concentration in pre-industrial times.[72] Projections reported in the Second Assessment Report of the IPCC have shown that stabilising CO_2 concentration at 550 ppmv by the end of the next century would require global emissions ultimately to fall to well below their current levels (IPCC, 1996).

The question arises of whether the targets agreed upon in Kyoto are in line with a stabilisation path to keep concentration below 550 ppmv or, as an extreme opposite position, whether effectively this target is already out of reach. A study by the IPCC (Wigley *et al.*, 1997) has shown that limitations of the magnitude decided in Kyoto[73] "would [not] lead to anything approaching CO_2 concentration stabilisation" with unconstrained growth of emissions in non-Annex 1 countries. The study concludes that "substantial global emissions reductions beyond those defined by the various emissions limitation proposals would be required". While this is not in dispute, several emission trajectories may lead to this outcome. On an optimistic note, Wigley (1997) argues that non-Annex 1 countries can wait several decades before their emissions need to depart significantly from their BaU trajectories, even in the context of stabilisation at 550 ppmv. Such optimism crucially relies on the assumption that world emissions can ultimately be reduced to a very low level (see Annex 2). In addition, these studies are based on a central BaU scenario (IS92a) which looks optimistic in the light of recent high emissions growth.

Linking past emissions to future concentrations is subject to a number of important uncertainties. First, the carbon cycle has been described by different scientific models which imply different profiles of carbon decay in the atmosphere. However, the sensitivity analysis made in Annex 2 indicates that the choice of model has little impact on the results. Second, the predictive value of these models may fall for concentrations departing significantly from those observed in the past for the reasons described in Chapter 2.

These uncertainties need to be borne in mind when interpreting the current attempt to translate CO_2 emissions, as projected by the GREEN model, into future concentration levels (see Annex 2).[74] The results of this exercise roughly confirm the previous analysis made by the IPCC prior to the Kyoto meeting. As concerns action by Annex 1 countries alone, they can be summarised as follows:

- The CO_2 limits on Annex 1 countries set by the Kyoto Protocol, if extended beyond the first commitment period, will do little to stabilise CO_2 concentration (at least over the time horizon 1997-2200). At most, the targets decided in Kyoto might delay by about a decade the point in time when the 550 ppmv level would be reached.
- Any further effort by Annex 1 countries alone to reduce their emissions below the levels decided in Kyoto will have little impact on concentration, given their already rapidly declining share in global emissions.

The costs of "missing" Kyoto

These results could be interpreted as putting the merit of the Kyoto Protocol in doubt. Indeed, *a priori*, there is no guarantee that the Kyoto Protocol will be the first step of a cost-effective path towards a stable concentration level. In choosing the least-cost pathway, one has to balance the costs of early and more gradual action against the costs of later, more rapid, forced action. Reasons for delaying action involve: *i)* to give more time for the economic turnover of existing capacities; *ii)* to give more time to develop low-cost alternative carbon-free energy sources; and, *iii)* the effect of time discounting. Based on "integrated assessment" models, some authors have argued that the least costly way to achieve concentration stabilisation would be to let emissions continue unconstrained for a couple of decades, followed by drastic cuts later on (Wigley *et al.*, 1996). In that sense, the reductions specified in the Kyoto Protocol would not be consistent with a cost-effective strategy to stabilise concentration (Manne and Richels, 1998).

However, delayed action involves the risk that, as scientific information becomes available, more rapid reductions will become necessary, causing a premature retirement of future capacity. By explicitly taking account of both uncertainty and inertia in changing the productive system which generates GHGs, Ha-Duong *et al.* (1997) reach the conclusion that a strategy of early and modest abatements may prove less costly if there is a high risk of exceeding the stabilisation target or "if the BaU scenario is significantly higher over the coming years than the IS92a scenario used by the IPCC", as the projections from the GREEN model would suggest.

Both approaches mentioned above fail to take into account a number of additional factors. First, they neglect the role of early abatement in stimulating the process of learning-by-doing which, in turn, would lower subsequent abatement costs. Second, they do not take into account the equity aspects of the international negotiation process, which make non-Annex 1 countries reluctant to undertake emission reductions before Annex 1 countries have taken some action (see below). These arguments reinforce the view that early but moderate reductions, as those specified in the Protocol, are the best way towards an adequate response to the threat of climate change, given the uncertainty and the inertia which characterise both the energy system and the negotiation process.

4.2. Establishing a global agreement

Analysis of emission and corresponding concentration timepaths in Annex 2 establishes that to stabilise CO_2 concentrations, or even to delay significantly the date by which they reach double the pre-industrial level, will require the Kyoto Protocol to be extended after the first budget period to involve non-Annex 1 countries in a substantial mitigation effort. This section deals with the requirements for obtaining participation by non-Annex I countries while the next section considers the implied economic costs.

The Kyoto Protocol was agreed on a voluntary basis. The countries which made commitments may be seen as having weighed the gains of being a free rider against the risk that no agreement would be struck. In this sense, the rationality which underlies the process of climate negotiation may be seen as at least partly related to economic incentives. Against this background, a growing game-theoretic literature analyses the formation of coalitions for environmental objectives such as to combat climate change. Box 6 reviews the main outcomes from this approach. In summary, these studies agree that a fully voluntary agreement would comprise at most a small number of signatories and, therefore, be insufficient to stabilise the climate. A main reason for this pessimistic conclusion is the strong incentive for countries to free-ride. The literature also suggests that if carbon leakages are significant, it may be even more difficult to reach wide agreement because incentives to free-ride become stronger. Conversely, the existence of local, ancillary benefits may strengthen the incentives of countries to participate in an agreement. Studies have also identified a number of possibilities to expand this small "self-enforcing" coalition, including by means of transfers to hesitant countries.

It may be questioned whether simple theoretical models focusing on economic incentives have much predictive value as regards the current process of climate change negotiation.[75] The models are generally not good at taking into account the asymmetric nature of climate change negotiations. A few studies have attempted to validate the results of the theoretical models in the case of heterogeneous participants (for instance, Botteon and Carraro, 1998). Although they seem to confirm the results from the simple model with identical countries, these approaches so far lack generality and hardly lead to any policy recommendation. Asymmetries are related not least to the fact that the impacts of climate change are likely to be very different across countries (see Chapter 5). At the same time, countries' preference for and valuation of climate change, and actions against it, may differ. Thus, it is often argued that environmental protection may be considered a superior good. Following from this line of reasoning, the countries which made commitments under the Protocol might be thought to have a high preference for a clean environment but, on the other hand, to suffer relatively little from the consequences of climate change (in some cases, perhaps even to benefit). In contrast, developing countries would be more affected by climate change but might be relatively less concerned about this. The current stage of the negotiation process would, according to this view, reflect the low level of expected damages in industrialised countries, together with the low preference for climate stability in the developing world. As environmental concerns in the developing world increase together with real income levels, the current agreement would then gradually expand. However, this process might be slow and could, accordingly,

Box 6. Establishing a global agreement – the theory

Game theory identifies the conditions for a coalition of countries taking policy action with a global environmental objective to exist and be stable: first, each participant in the coalition should gain compared with the initial situation where nobody co-operates; second, nobody should have an incentive to leave or to join the coalition. With these two conditions satisfied, the coalition is said to be "self-enforcing".

The problem of forming a self-enforcing coalition can be illustrated by comparing the net benefits of participant and non-participant countries under the assumption that all countries are symmetric (*i.e.* they have identical marginal abatement costs, damages and environmental preferences). Countries outside the coalition enjoy the global environmental benefit from the action undertaken by the co-operating countries without supporting any abatement costs. As a result, countries have an incentive to free-ride. This is illustrated in Panel A of Figure 11 where the payoff of non-participants (*i.e.* their net welfare gain) always exceeds that of coalition countries whatever the size of the coalition. Payoffs increase monotonically with the size of the coalition, as the environmental gains rise with the number of participants. The figure also illustrates that there may be an equilibrium at which no country has an incentive to join the coalition (because its payoff outside the coalition is higher than the payoff it would have as a participant in an enlarged coalition) and no country has an incentive to leave the coalition (because its payoff in the coalition is higher than the payoff it would have outside a smaller coalition).[1] This equilibrium determines the size of the self-enforcing coalition. It has been demonstrated that this size is typically small (Hoel, 1991; Barrett, 1992, 1994, 1997 and Carraro and Siniscalco, 1992). Botteon and Carraro (1998) have shown that this conclusion may hold when some types of asymmetries across countries are accounted for.

In the context of climate change, an important factor in determining the size of a self-enforcing coalition is the magnitude of carbon leakages. High carbon leakages reduce the global environmental benefit so that coalition countries would pay the abatement costs without receiving the corresponding benefits, thus reducing the incentive to form a coalition (with high leakage the coalition payoff may no longer be monotonically increasing as illustrated in Panel B of Figure 11). At the same time, non-participating countries get a higher payoff associated with the economic benefits of carbon leakages (lower energy prices, improved competitiveness of energy-intensive industries). Under these conditions, the size of the self-enforcing coalition is reduced or there may even be no incentive at all to form any coalition (see, for instance, Carraro and Moriconi, 1998). Here again, this conclusion may remain valid when countries are not identical[2] (Botteon and Carraro, 1998).

Action to reduce greenhouse gas emissions may have important local, ancillary benefits. These benefits will not accrue to free-riders, thereby reducing the incentives for free-riding and increasing the size of the self-enforcing coalition. Whether in practice the effects from leakage or the effects from local ancillary benefits will dominate is unclear.

Box 6. **Establishing a global agreement – the theory** *(cont.)*

Figure 11. **The theory of coalition formation: payoff curves at different coalition sizes**

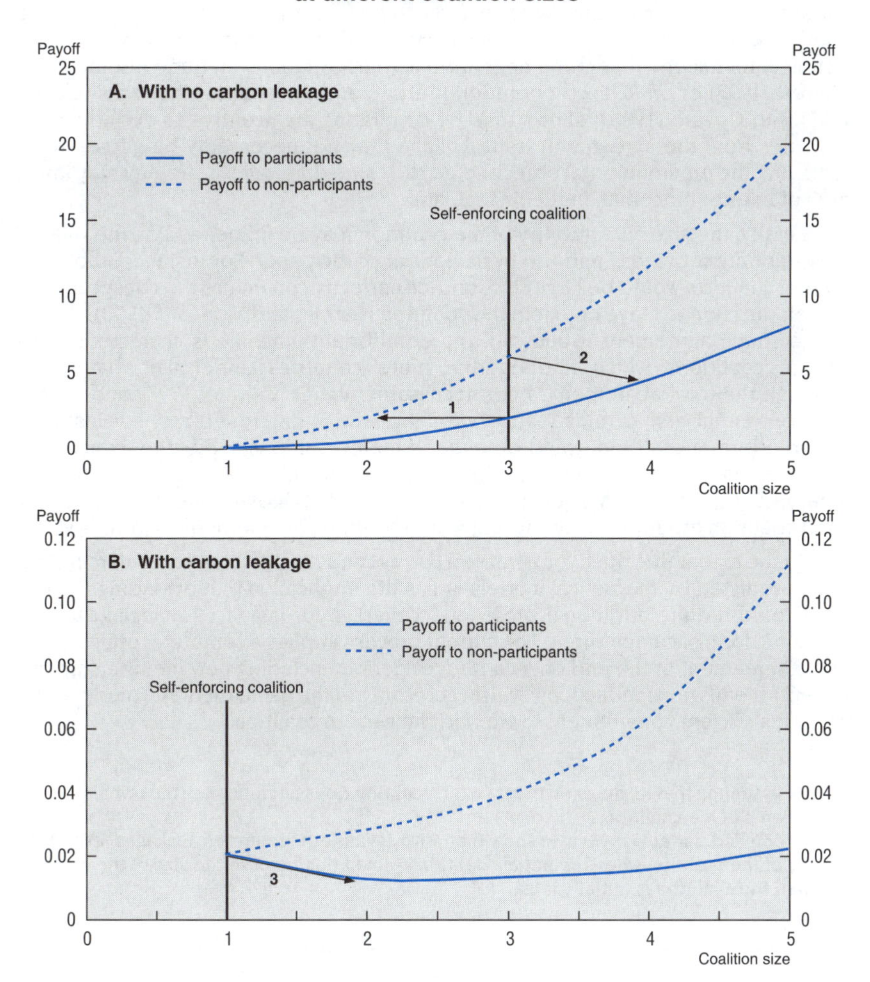

1. Internal stability: no participant has incentive to leave the coalition.
2. External stability: no non-participant has incentive to join the coalition.
3. Nobody has any incentive to form a coalition; the equilibrium coalition is a singleton.

Source: Based on Carraro and Moriconi, 1998.

Box 6. Establishing a global agreement – the theory (*cont.*)

The size of an environmental coalition can be expanded in a number of ways. First, a system of transfers may be designed so that no country refuses to join a coalition. However, for such an outcome to be achieved, a group of "core" countries need to be committed to co-operation (Carraro and Siniscalco, 1993; Hoel, 1991; Carraro, 1998*a*). In that sense, the resulting agreement is no longer self-enforcing. Another way to expand a coalition is by "issue linkage": it has been suggested, for instance, to link environmental negotiations to negotiations on trade liberalisation (Barrett, 1995) or on R&D co-operation (Carraro and Siniscalco, 1995; Katsoulacos, 1997). But Carraro (1998*a*) shows that, by generating an incentive to exclude some countries from the agreement, issue linkage may not necessarily be successful in achieving the environmental objective. At the same time, such a strategy may imply risks of inferior outcomes in the linkage area.

Finally, the size and stability of the coalition may be influenced by the rules of the negotiation process and the behaviour of participants. For instance, the equilibrium coalition would be larger if countries correctly conjecture the consequences of their decision to leave or to join the coalition (Carraro and Moriconi, 1998). If more than a single agreement is allowed, the equilibrium outcome is characterised by several coalitions which implies that more countries undertake abatements although the total abatement is not necessarily higher (Carraro, 1998*a* and 1998*b*). Different membership rules leads to coalitions with different sizes. For instance, the equilibrium coalition would be larger if membership is subject to a unanimity rule – *i.e.* that no entry or exit is allowed without unanimous consent of coalition members (Carraro and Moriconi, 1998). Here again, achieving the grand coalition (*i.e.* a coalition grouping all countries) would require some form of commitment.

To the extent that the Kyoto Protocol corresponds to the small self-enforcing coalition predicted by theoretical models, it has the implication that extending the Protocol could require additional provisions, covering, for instance, financial transfers and some form of commitment. The current process implies a complete renegotiation of the agreement at the end of each budget period, including new membership and targets for emissions reductions, which, according to the game theoretic models, may imply insufficient commitment to expand the current coalition.

1. And assuming that its decision to leave the coalition does not influence the remaining participants in the coalition.
2. Botteon and Carraro (1998) also show that, with asymmetric countries, leakages reduce the size of the coalition when the initial coalition is small but increase the size of the coalition when the coalition is initially large.

result in an eventual very high level of concentrations. These considerations obviously take into account neither uncertainty nor the profile of risk, with some low probability but disastrous outcomes.

More fundamentally, there is a question as to how far economic rationality is useful in describing coalition formation and the future of the negotiation process. Insofar as anthropogenic emissions of greenhouse gases have contributed to increased atmospheric concentration levels in the past, responsibility lies largely with the Annex I countries, whose emissions represented two-thirds of the total as late as in 1995. Against this background, and emphasising basic principles of fairness, most developing countries have been reluctant to accept restrictions on their own emissions, which are generally far below those in OECD countries measured on a *per capita* basis, at the same time as OECD countries are far richer than developing countries. Indeed, practically any scheme for reducing global emissions induces changes in the international distribution of income. In the context of tradable emission rights this effect is immediately apparent – recipients of large quantities of emission allowances appear to be receiving transfers from those with little or no allocation. It is useful to consider this allocation issue from three different perspectives on property rights.

A first perspective is provided by considering the property rights that are implicit in the UNFCCC. The convention notes, *inter alia*, that climate change is a common concern of mankind, that efforts to deal with it should be based on a principle of equity, and that the Parties to the convention are determined to protect the climate for both current and future generations. One might conclude from this that since the climate is a common concern, it is a common good, and that equity implies that each individual has an equal stake in it. From there it is a small step to argue that a neutral allocation is to give each individual an equal right to greenhouse gas emission permits – equal rights to damage the climate.

Another perspective would be to consider the consequences of regime changes. Up to now, there has been no legal notion of greenhouse gas emissions as a bad[76] and economic agents have not been expected to act as if they were. Any legal change which discriminates on the grounds of past emissions may be inequitable in the sense of being a retrospective penalty. Taking a step further, it might then be argued that equity requires grandfathering of past emission patterns.

A third perspective stresses equity from an economic point of view (rather than as having to do with physical emission rights). This perspective emphasises income distribution from the point of view that those with the broadest shoulders should carry the heaviest burdens. In this case, it may be argued that allocation of emission rights should reflect relative incomes.

There is no obvious way to choose between these three interpretations. In practical terms, what is needed is a way of allocating emission allowances which encourages both current and future heavy emitters to participate. One way to do this might be to have a transition through time, for example, from allocations based on grandfathering to equal *per capita* allocations.[77] Consensus on such an allocation scheme might induce non-Annex 1 countries to adopt voluntary tradable emissions limits for the first commitment period, without weakening the overall constraint on emissions imposed by the Kyoto Protocol – no "hot air" would be generated. It would avoid penalising Annex 1 countries so heavily that they would be likely to reject the whole process. It may not, on the other hand, give non-Annex 1 countries sufficient incentive to participate.

This review does not offer any clear indications how to bring non-Annex 1 countries to agree to emission constraints in the near future beyond certain generalities: reliance on perceived self-interest will be insufficient; some notion of equity will need to be satisfied; some financial transfers will probably be necessary to bridge the gaps between different ideas of what of equity implies. The next section considers some of these issues and attempts to quantify their importance.

4.3. *Economic costs of stabilising* CO_2 *concentration*

This section presents some illustrative quantifications of the costs to Annex 1 countries of global agreements to achieve stabilisation of concentration levels, including the role of financial compensation to induce non-Annex 1 countries to participate.

A variety of emissions pathways may lead to stable concentration levels by the end of the 22nd century. Here, however, the focus is on a shorter horizon – from the first commitment period (2008-12) to 2050. Accordingly, the scenarios analysed below may be characterised better in terms of different degrees of risk aversion rather than stabilisation *per se*. The large uncertainties involved in "translating" emission paths to concentration levels also points to an emphasis on risk aversion rather than targeting of concentration levels. The analysis considers three global emission scenarios which imply different degrees of emission reductions but have in common that participation by non-Annex 1 countries is required to achieve the limitations:

- A first scenario is based on the premise of low risk aversion: emission reductions are phased in so as to let concentration rise towards 550 ppmv of CO_2 in 2080 (the "low-cost 550 ppmv" scenario in Figure 12) and then broadly stabilise at that level.
- Alternatively, a scenario corresponding to high risk aversion would imply more drastic emission reductions in order to keep concentration well below 550 ppmv over the next century (the "high-cost 550 ppmv" scenario in Figure 12).
- In the third scenario, concentration rises to 740 ppmv with stabilisation achieved only by the middle of the 22nd century (the "740 ppmv" scenario in Figure 12).

Figure 12. **The long term: alternative CO_2 concentration pathways**

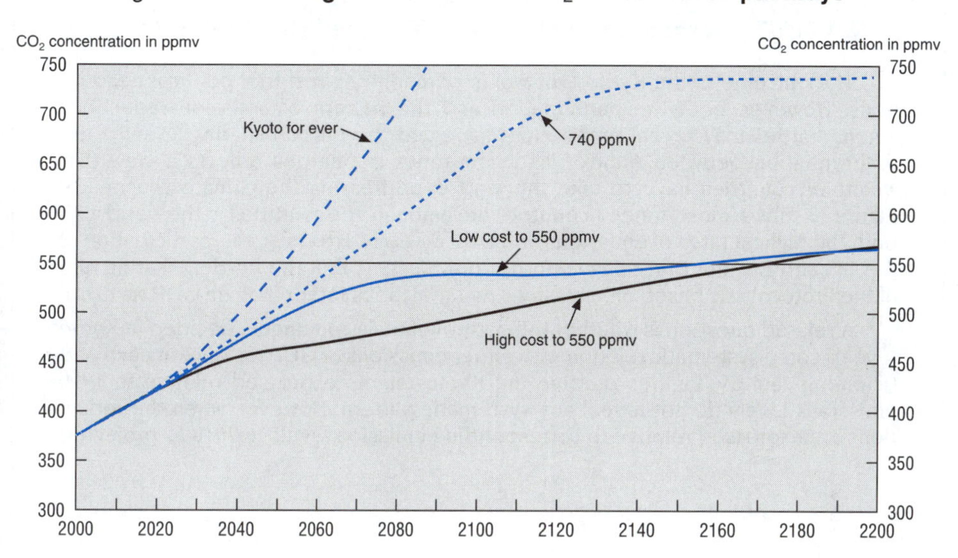

Source: GREEN Model, OECD Secretariat.

Burden-sharing

There is also a wide variety of ways to share the burden of these reductions between Annex 1 and non-Annex 1 countries. In principle, the Kyoto targets imply a burden-sharing agreement and one option might be to extend any equity rule underlying the current version of the Protocol to non-Annex 1 countries. In practice, however, the burden-sharing underlying the Protocol is not easy to identify with any simple rule (Box 7). In particular, it is unclear whether the undertaking and the magnitude of Parties' commitments are based on some criterion related to emissions *per capita* or income *per capita* or, alternatively, can be best seen as an outcome of the negotiation process. Furthermore, the Protocol does not provide any guidance about the principle underlying the setting of emissions target beyond 2012. Given this uncertainty, the analysis of possible extensions of the Protocol is based on a few archetypal rules which may be seen as benchmarks rather than specific proposals.

One benchmark equity rule is based on the principle of "ability to pay": *i.e.* that non-Annex 1 countries accept commitments when their real income *per capita* reaches a given threshold and that commitments are proportionate to relative

Box 7. Is there an equity rule behind the Kyoto Protocol?

The outcome of the Kyoto Protocol is primarily the result of political negotiations. However, both the participation and the pattern of emission reductions among Parties may reveal preferences as regard the burden-sharing. Clearly, the differentiation between Annex 1 and non-Annex 1 countries reflects a view that wealthier countries have to take the lead in addressing the climate change. As Figure 13 shows, most Annex 1 countries are amongst the countries in the world with both the highest rates of emissions and GDP *per capita*. However, the participation of the CIS and Eastern European countries may suggest that the burden-sharing rule of the Protocol was based on emissions *per capita* ratios rather than on GDP *per capita*.

A related question is whether the pattern of emission targets decided in Kyoto reveals some systematic relation with either emissions or GDP *per capita* in each participating country. Figures plotting the Kyoto targets expressed relative to 1990 emissions levels do not reveal any systematic pattern. However, when the reductions are expressed relative to corresponding emission levels in 2010 as projected

Figure 13. **Emissions and GDP per capita in 1995**

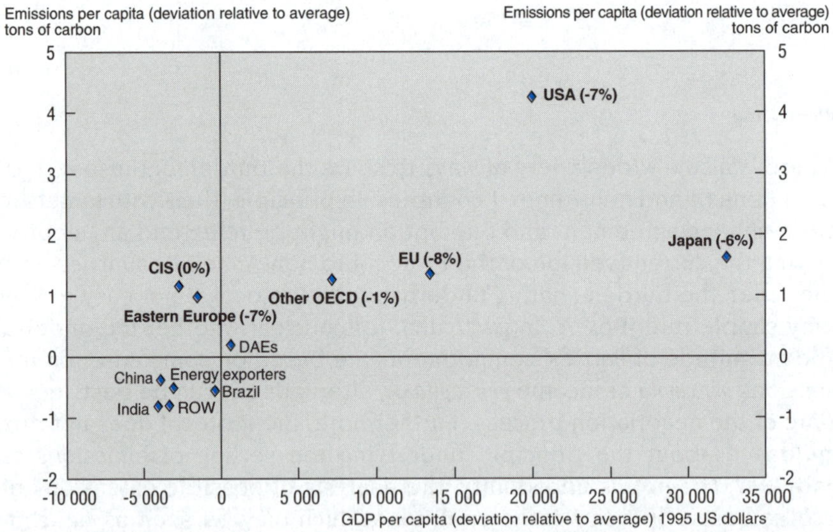

Note: Percentages in brackets correspond to the emission reductions of the Kyoto Protocol (relative to baseline).

Source: GREEN Model, OECD Secretariat.

Box 7. Is there an equity rule behind the Kyoto Protocol? (*cont.*)

by the IPCC, a correlation can be identified, although with a low degree of significance (see Figure 14). This correlation indicates that Annex 1 countries with a higher level of GDP *per capita* have tended to accept higher emissions reductions in 2010 relative to baseline levels.* According to the logarithmic fit on Figure 14, a doubling of the GDP *per capita* would correspond to a 14 per cent increase of the willingness to reduce emissions in 2010. Some countries - like the United States, Canada, Japan and some southern European countries - have committed to abatements in excess of this implicit "ability to pay" rule while other European countries, the CIS and some Eastern European countries have been relatively less ambitious. These results suggest that, if any systematic pattern can be found in the outcome of Kyoto, the negotiation process has considered the commitments in terms of 2010 abatements, thus taking into account the unconstrained growth of the emissions up to the first budget period, as projected in the baseline scenario, rather than the targets *per se* expressed relatively to 1990 levels of emissions.

Figure 14. Emission reductions in 2010 versus GDP per capita for Annex 1 Parties

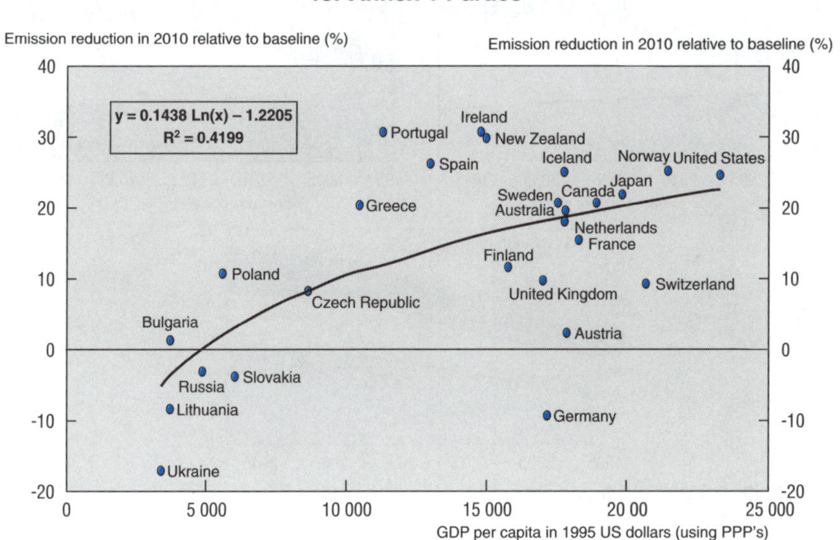

Source: IEA and IPCC.

* No positive correlation was identified between emissions reductions and the level of emissions par capita in each Annex 1 Party. For EU countries, reductions are based on IPCC projections and the aggregate EU reduction rate relative to 1990 as specified in the Kyoto Protocol (–8 per cent).

Figure 15. **Burden-sharing in alternative emission scenarios, 2010-2050**

Annual emission levels

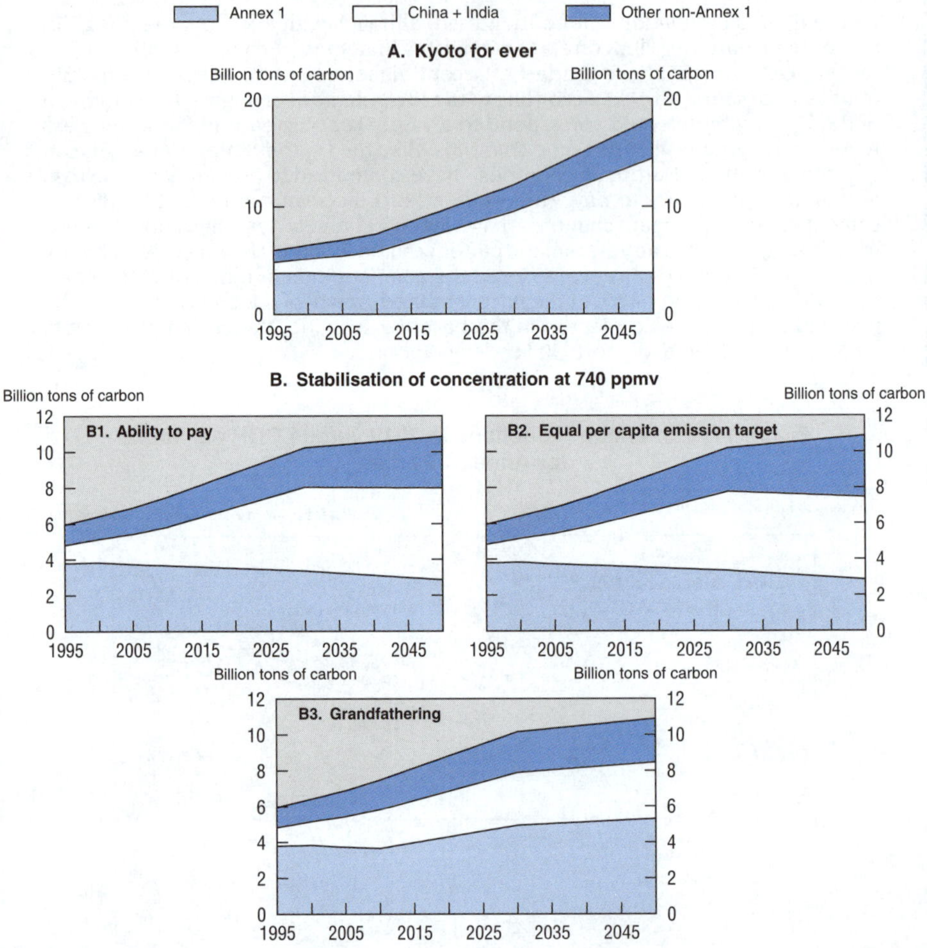

Source: GREEN Model, OECD Secretariat.

income *per capita* levels (see Annex 3). Alternatively, CO_2 emissions *per capita* can be considered as the criterion for sharing the burden of future abatements. Following this equity rule – subsequently referred to as an "equal *per capita* emissions" rule – non-Annex 1 countries join the coalition to reduce emission, *i.e.* become subject to constraints, when their emissions *per capita* reach the same level as those in Annex 1

Figure 15. **Burden-sharing in alternative emission scenarios, 2010-2050** *(cont.)*

Annual emission levels

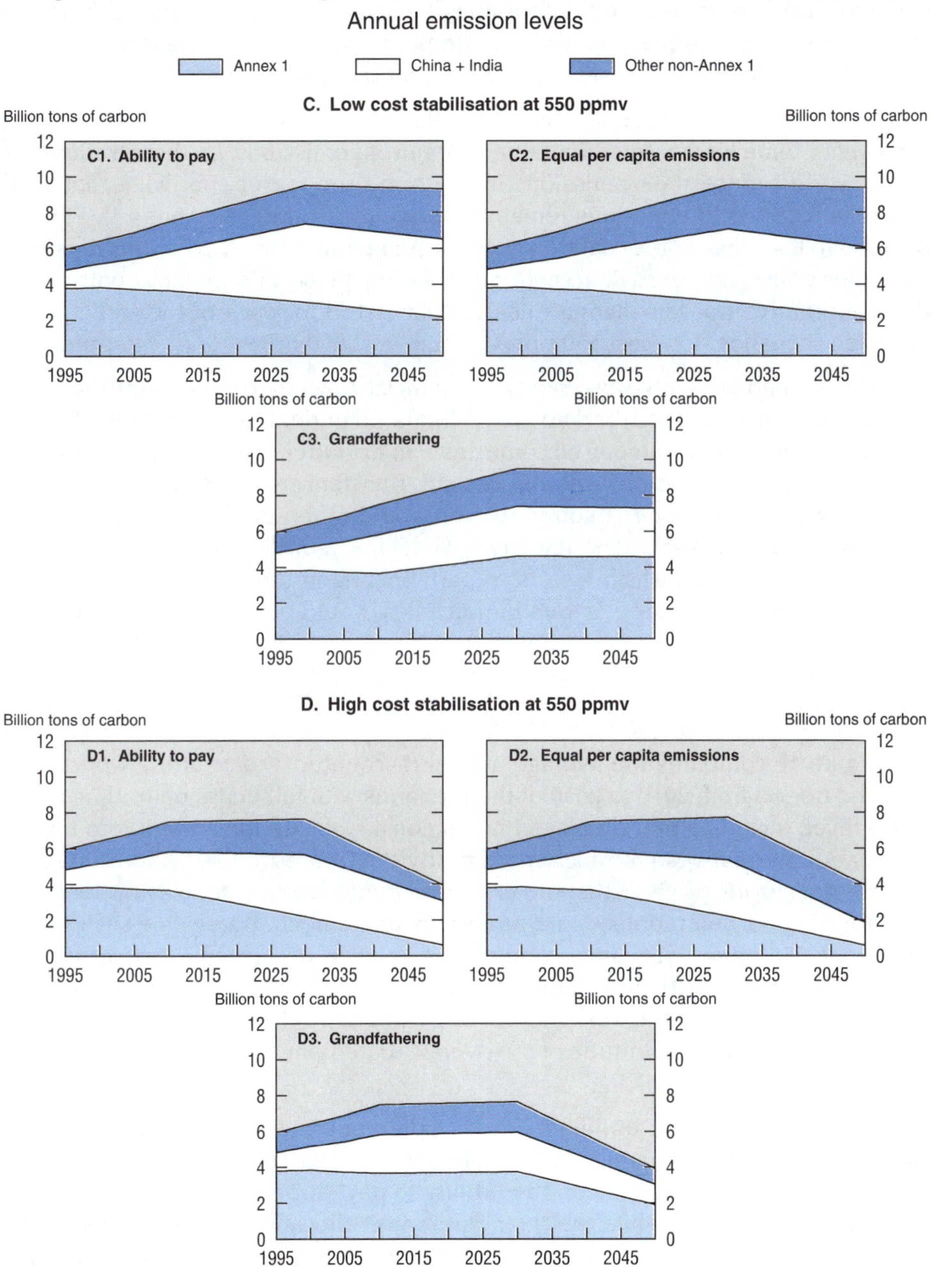

Source: GREEN Model, OECD Secretariat.

countries (see Annex 3). In this case, all countries in the world would ultimately converge towards the same level of emissions *per capita*. As Figure 15 shows, both the "ability to pay" and "equal *per capita* emissions" rules imply an increase in the share of non-Annex 1 countries in total emissions over the longer term.

A very different burden-sharing rule is that allocation of future emission rights represents some kind of "grandfathering" of current emissions, *i.e.* that all countries in the world reduce their emissions in the same proportion, implying that their shares in total world emissions remain constant over time (see Annex 3). For this position to lead to a global coalition would require non-Annex 1 countries to accept that their emissions *per capita* remain much below those in Annex 1 countries. All three alternative burden-sharing rules are simulated for each of the three global emission scenarios "low-cost 550 ppmv", "high-cost 550 ppmv" and "740 ppmv".

The scenarios involve two types of financial flows between countries. First, each of the scenarios described above is simulated under the assumption of either no or full permit trading among all countries.[78] In the latter case, financial flows arise as counterparts of emissions trading. Second, transfers are used to induce participation by those non-Annex 1 countries who would otherwise have an incentive to free-ride: each scenario is simulated with the additional assumption that non-Annex 1 countries which lose from participating in the common action, even taking account of the proceeds from permit sales, would be compensated by financial transfers from Annex 1 countries.[79]

Economic costs

Figure 16 compares the average annual discounted[80] costs at the world level over the period from 2010 to 2050 of the scenarios with full emission trading at the world level, including that discussed in Section 4.1 of extending the Kyoto targets forever. All alternatives are much more costly than the extended Kyoto Protocol, which is not surprising given that the extended Kyoto Protocol comes nowhere near to stabilising concentrations. More ambitious abatement strategies – such as the "high-cost 550 ppmv" scenario – would cost average real income loss of around 0.6 to 0.8 per cent. With perfect emission trading among all countries, in principle, the aggregate costs for the world do not depend on the choice of the burden-sharing rule.[81] Rather, the magnitude of the overall abatement effort is what determines the aggregate costs.

Assuming no permit trading leads to a different outcome (Figure 17). In this case, different burden-sharing rules lead to different aggregate world costs. With no permit trading, the rule based on the "ability to pay" appears as the most costly in all abatements options while the "grandfathering" allocation rule is cheapest. The rule based on an "equal *per capita* emissions target" has an intermediate ranking. These differences relate to the fact that each rule implies a different allocation of

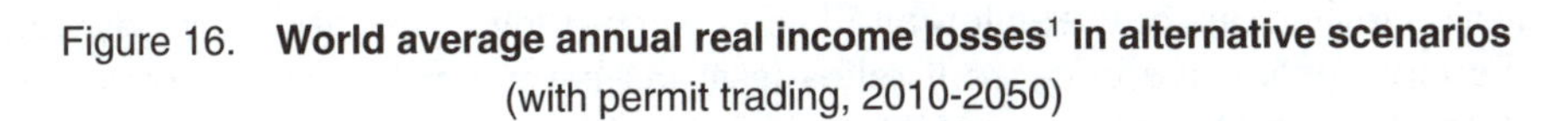

Figure 16. **World average annual real income losses[1] in alternative scenarios**
(with permit trading, 2010-2050)

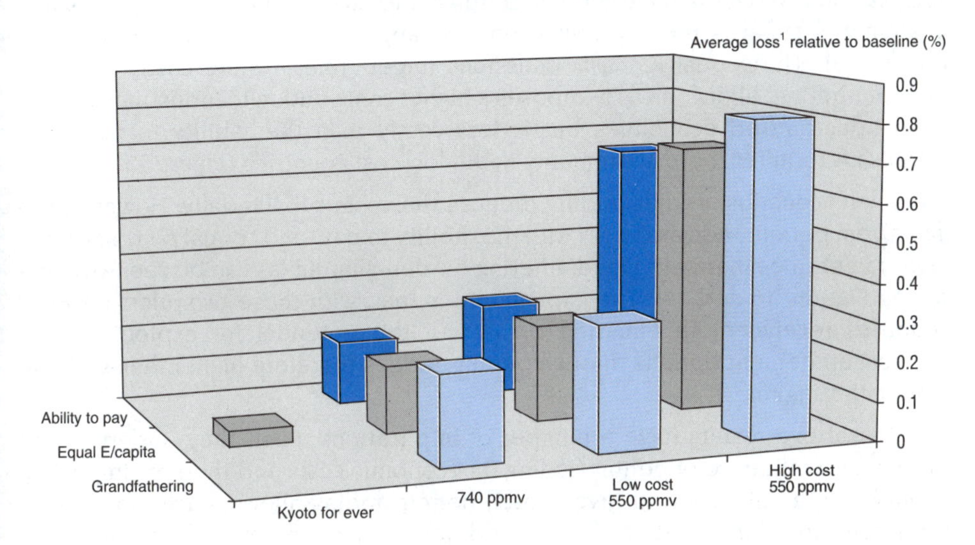

Figure 17. **World average annual real income losses[1] in alternative scenarios**
(without permit trading, 2010-2050)

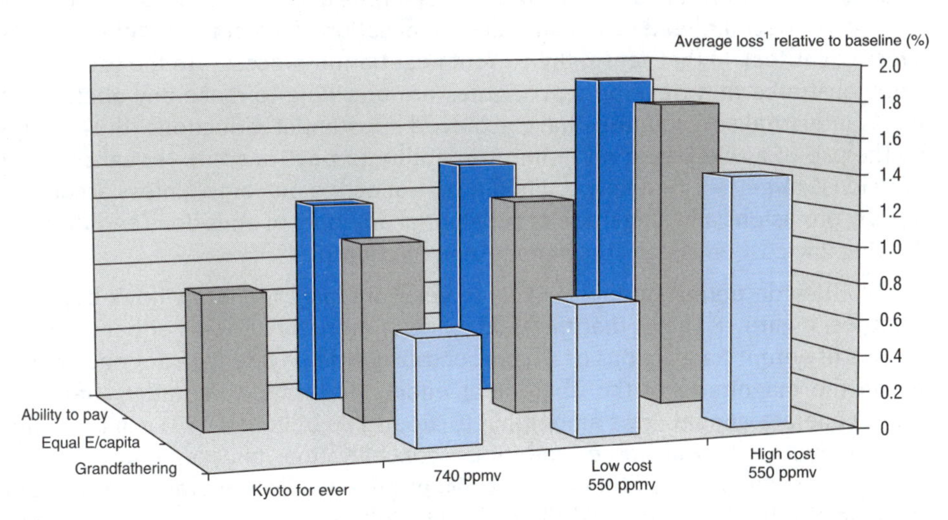

1. In terms of equivalent-variation of household real income, with a 3% discount rate.
Source: GREEN Model, OECD Secretariat.

emission reductions between low and high-cost countries. The "ability to pay" rule is the most expensive because it relies less on abatement in low-cost countries, such as China and India (Figure 15). In contrast, the "grandfathering" rule, as it shifts abatements towards the low-cost countries, appears globally much less costly (Figure 15). The "equal *per capita* emissions target" rule is more costly than the "grandfathering" allocation as it imposes a higher contribution to industrialised and semi-industrialised economies but is less costly than the "ability to pay" rule because it requires more abatements in the low cost countries (Figure 15).

In all scenarios, permit trading reduces the costs substantially. However, the gains from permit trade are larger with the "ability to pay" and "equal *per capita* emission" target rules than with grandfathering burden-sharing (as can be seen by comparing Figures 16 and 17). The explanation is that, with these two rules, low-cost countries receive more emission rights and the potential for exploiting their low-cost abatement options through permit trade is therefore higher than with the grandfathering rule.

The above results have a number of implications. First, they underline the longer-term influence of permit trading on economic costs and thereby, implicitly, environmental efficiency. At given costs, permit trading allows much more ambitious concentration targets to be reached: for instance, by making the option of early steps toward stabilising concentration at 550 ppmv less costly than the scenario "Kyoto for ever" without permit trading, despite the latter having almost no impact on concentration. Second, the results show that the way future emission reductions or permits are allocated will matter for the aggregate costs unless permit trading is perfect (*i.e.* leading to complete equalisation of marginal abatement costs across countries). In that sense, the current negotiation process – to the extent that it demonstrates that developing countries are unwilling to make any abatement unless industrialised countries have achieved substantial reductions (thus implying the use of a rule based on either the "ability to pay" or some "equal *per capita* emissions" target) – is unlikely to yield a cost-effective outcome, unless accompanied by provisions allowing trade of substantial amounts of emissions and thereby allowing low-cost emission cuts in non-Annex 1 countries.

World-wide permit trading is likely to generate large monetary flows between countries. Figure 18 shows that the total value of monetary flows – either as counterparts of permit transactions or as compensatory side payments – depend primarily on the magnitude of the abatement effort. Moderate abatement strategies would generate annual flows amounting to around 50 billion of 1995 dollars annually on average.[82] But more ambitious concentration targets – such as the "high-cost" scenario of stabilisation at 550 ppmv – would generate much larger flows, amounting to 150 to 200 billion of 1995 dollars annually. To put these estimates in perspective, the total net flow of official development assistance amounted to 48 billion dollars in 1997. Figure 18 also highlights the trade-off

Figure 18. **Average annual monetary flows[1] under alternative burden-sharing rules** (average over 2010-2050)

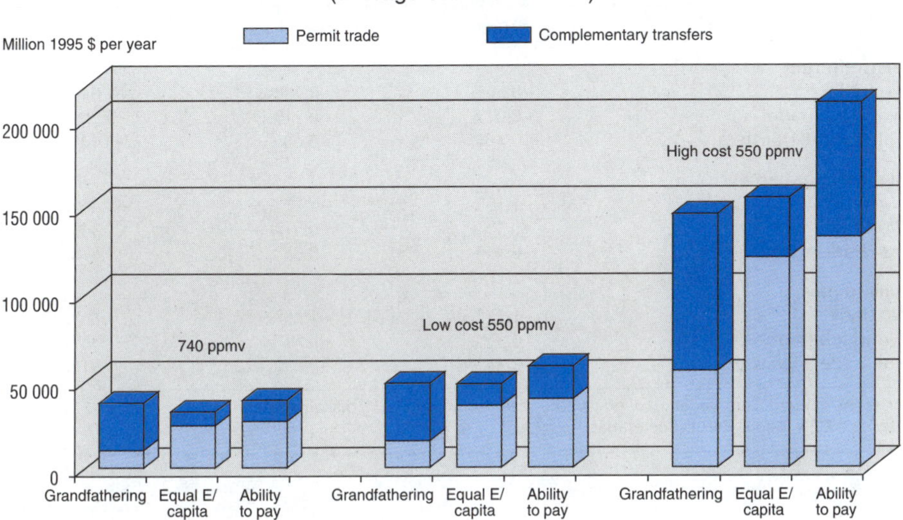

1. Average discounted flows, with a 3% discount rate.
Source: GREEN Model, OECD Secretariat.

between permit trading and side payments: the rules based on the "ability to pay" or "equal *per capita* emissions targets" generate more international redistribution through permit trading while the "grandfathering" rule requires a relatively higher proportion of side payments in order to generate sufficient incentives for the formation of a global coalition.

Table 10 shows the results for Annex 1 countries. In most cases, stabilising concentrations is relatively costly when Annex 1 countries are not allowed to trade (*i.e.* buy) emission rights. The lowest estimate is with the "grandfathering" rule, ranging from 0.6 to 1 per cent of real income per year, depending on the magnitude of the abatement. In contrast, real income losses in the scenario describing the high-cost pathway to 550 ppmv amount to 1 to 1.5 per cent per year. Furthermore, in this latter case, emission trading tends to have a lower cost-saving impact: first, because marginal abatement costs at such high levels of reductions tend to converge across countries to the price levels set by available backstop energy sources, thus reducing the potential efficiency gains from trading permits; second, due to the shape of the cost curves when backstop energy sources become available.[83]

Table 10 also reports the average costs to Annex 1 countries if they engage in transfer payments to compensate non-Annex 1 countries for any real income loss

Table 10. **Costs of alternative burden-sharing rules for Annex 1 countries**
Average % deviation[1] relative to baseline over 2010-2050

	740 ppmv	Low cost 550 ppmv	High cost 550 ppmv
Grandfathering			
No trade	–0.56%	–0.70%	–1.04%
World full trade	–0.21%	–0.31%	–0.76%
Full trade + transfers	–0.46%	–0.60%	–1.42%
Equal em. per capita			
No trade	–1.15%	–1.22%	–1.46%
World full trade	–0.35%	–0.51%	–1.34%
Full trade + transfers	–0.46%	–0.65%	–1.63%
Ability to pay			
No trade	–1.17%	–1.25%	–1.46%
World full trade	–0.35%	–0.51%	–1.36%
Full trade + transfers	–0.48%	–0.69%	–1.91%

1. In terms of equivalent-variation of household real income, with a 3% discount rate.
Source: GREEN Model, OECD Secretariat.

incurred from their participation in emission reductions.[84] In the scenario where concentrations are kept significantly below the 550 ppmv threshold during the next century (*i.e.* the "high-cost 550 ppmv" scenario), Annex 1 countries would suffer an average loss of around 1.4 to 2 per cent of real income annually.[85] This is costly given that it is an average estimate calculated over a long period of time (40 years) with a discount rate of 3 per cent. Losses of such magnitude clearly raise questions about the feasibility and opportunity of following such a strategy. Here again, the nature of the burden-sharing rule matters, with the rules based on the "ability to pay" or the "equal *per capita* emissions target" being substantially more costly than the "grandfathering" rule.

For non-Annex 1 countries, the real income effects range from losses of 3 per cent in the "high-cost 550 ppmv" scenario without trading to gains in scenarios involving financial compensations (Table 11).[86] In general, non-Annex 1 countries would benefit from trading permits to a larger extent than Annex 1 countries (in particular, in the most stringent "high-cost 550 ppmv" scenario). For non-Annex 1 countries too, the "ability to pay" and "equal *per capita* emissions" rules of burden-sharing paradoxically appear relatively more costly than grandfathering in the absence of tradable permits.[87] For both sets of countries, the variation in economic outcomes as a result of differences in burden-sharing is less than the variation due to different degrees of ambition – or risk aversion – concerning concentration profiles. In particular, the high risk-aversion scenario appears very costly, raising the question whether the benefits exceed the costs. The next chapter takes up the issue of costs of living with climate change.

Table 11. **Costs of alternative burden-sharing rules for Non-Annex 1 countries**
Average % deviation[1] relative to baseline over 2010-2050

	740 ppmv	Low cost 550 ppmv	High cost 550 ppmv
Grandfathering			
No trade	–0.65%	–0.78%	–2.13%
World full trade	–0.28%	–0.35%	–0.92%
Full trade + transfers	0.26%	0.27%	0.48%
Equal em. per capita			
No trade	–0.72%	–1.20%	–2.39%
World full trade	0.18%	0.29%	0.67%
Full trade + transfers	0.42%	0.59%	1.28%
Ability to pay			
No trade	–1.16%	–1.65%	–2.78%
World full trade	0.21%	0.33%	0.84%
Full trade + transfers	0.50%	0.70%	2.01%

1. In terms of equivalent-variation of household real income, with a 3% discount rate.
Source: GREEN Model, OECD Secretariat.

5. Adapting to climate change

According to consensus forecasts, some climate change is inevitable and is likely to continue over a long period, possibly several centuries. Although climate forecasting models are very complex, they give only relatively crude indications of how climates may change in specific regions. At the moment it is technically impossible therefore to run simulations to match a particular assumed path of greenhouse gas emissions with the consequent change in climate through time, even for broad regions of the globe, let alone individual OECD countries.

It is virtually impossible for countries to assess how their climate is going to change and already plan accordingly, even if somehow a convincing global agreement on the future path of emissions had been reached. There is consensus, however – as reflected in IPCC (1998) – on the likely nature of climate change and the consequent environmental impacts in different regions of the world, over the next 50 to 100 years (*i.e.* the short to medium term), under a particular set of assumptions about the effects of emissions. This chapter summarises that consensus. For a small number of countries national reports provide more detailed estimates. It then considers which economic sectors appear most sensitive to climate change impacts. The possible costs and benefits of these changes, and adaptation to them, are then discussed, along with the influence of uncertainty on how these issues are viewed.

5.1. *Anticipated climate change*

The following paragraphs are derived from the 1997 special IPCC report on the regional impacts of climate change (IPCC, 1998). That publication distinguishes ten regions, of which five do not contain any OECD country[88] and to which less attention is given here.[89] The regional reports do not all use precisely the same assumptions about global warming or sea-level rise, though they are broadly standardised on a doubling of the CO_2 concentration by the second half of next century, with a projected global temperature increase of 2.5°C (range 1.5-4°) and sea-level rise of 0.5 metres (0.2-1).

North America

North America covers a wide range of climatic types, and variability of the weather is already a factor. Climatic change in the 20th century has seen average temperatures peak in the 1940s before decreasing up to the 1970s, increasing again subsequently, being now around the level of the 1940s. Average precipitation gradually increased during the century, but there have been falls in some areas – mainly California and the Northern Rocky Mountain states. There has also been an overall trend towards an increasing proportion of precipitation from extreme events, though not in Canada. Although the sea level has been rising in general, it has been falling in some areas (for the most part owing to post-glacial rebound).

The IPCC reports that forecasts for average temperatures vary quite considerably, while there is more consensus on precipitation. A doubling of CO_2 concentration will increase winter temperatures by more than summer ones, but the range of forecasts is quite wide (1.4-4.8°C for the winter and 0-2.6°C for the summer). A further increase in the proportion of precipitation that comes from extreme events is expected, though there is disagreement on whether this would imply more storms, or more intense ones. The existing regional variation in sea-level rise is expected to continue, with the level rising fastest in the Gulf of Mexico and the mid and southern Atlantic coast.

Models can give only broad indications as to the potential, let alone the likely, response to forecast trends. The suitable area for forests will generally increase, but there is a big difference between the suitable area and actual forest cover because of human intervention. Forest types would generally need to migrate northwards. This may not happen easily because of gaps in forest cover and also because the movement of suitable zones is likely to be faster than many species could actually migrate naturally; on the other hand, human intervention could allow faster migration. Tundra and peatland will migrate northwards too, with the affected regions perhaps changing from being a net sink to a net source of carbon. Ice-jam patterns will change, generally so as to reduce frequency or intensity, but with possible increases in some areas.

Overall, agricultural productivity in North America would benefit from this projected climate change: more CO_2 and warming with more rain gives higher output, and there would be some gain even with moderately less rain; there may be some offsetting costs from possible increased variability, and from more frequent or more intense extreme events, however. Some areas, including Mexico, where water shortages already constrain agricultural output and where precipitation may fall, may see lower productivity, even before accounting for extreme events.

A 0.5 metre rise in sea level is estimated to threaten[90] between 8 500 and 19 000 km^2 of dry land in the United States and Canada, of which between a fifth and a quarter is developed land. The same rise would result in a net loss of between 17 and 43 per cent of coastal wetlands. River systems may be significantly affected: the possibility of lower summer precipitation, and the probability of higher evapo-transpiration, will make river transport more difficult to manage, and there may be more frequent or more severe floods.

Temperate Asia

This region includes China; little specific information on Japan and Korea is presented. The average temperature in the region has risen 1°C over a century, though not uniformly; this is due largely to higher winter temperatures, with only very slight increases in the other seasons. The increase has been faster since the 1970s. As for average precipitation, it has risen in Korea, but fallen in Japan (over a century, +10 to +20 per cent and –10 to –20 per cent respectively).

The regional mean temperature is expected to increase by between 1° and 3.5°C by 2100, but confidence in the sub-regional forecasts is very low. An important source of uncertainty is the influence of aerosols (deriving largely from sulphur-related emissions); some simulations imply summer cooling in parts of China. The importance of the monsoon, of typhoons and of the *El Niño/La Niña* phenomenon adds much uncertainty to precipitation forecasts; although a fall is generally expected, rises may also occur.

In Russia the tundra zone would decrease significantly with related expansion of grass and shrublands, as well as northern migration of forests in both Russia and China. The impact on agricultural productivity is uncertain, yields may fall without adaptation, especially if water shortages materialise as they may in China, but these may be more than offset by the carbon fertilisation effect.

Sea-level rises may have important consequences in Japan. The IPCC report notes that three major economic centres of Japan – Tokyo, Osaka and Nagoya – have large low-lying coastal areas. Two million people currently live below mean high-water level. If the sea level rose one metre (the upper end of the range usually quoted for the second half of the 21st century) this number would rise to 4.1 million; the same rise would increase the flood-prone area to 8 900 km^2 from 6 270 currently.[91] According to the report, a 30 cm rise in sea level would eliminate more than half of Japan's sandy beaches, and a one metre rise would eliminate nearly all of them; over the past 70 years, about one-third, measured by area, have disappeared. Significant parts of the Chinese coastline are also threatened.

On extreme events, while they may increase in the future (the report notes no strong evidence for such an increase in temperate Asia) deaths from typhoons, for example, have been decreasing since the Second World War, as a result of both mitigation and warning measures.

Australasia

This region covers Australia, New Zealand, and their outlying islands. It may be noted that in the current climate the variability of Australia's rainfall (especially in the north) is among the greatest in the world. The frequency and duration of drought (with flooding at other extreme) is great. To the extent that there is a tendency for other OECD countries' climates to move in this direction, Australia may be in a position to provide useful lessons on mitigation and adaptation to other countries.

Australasia has been warming a bit faster than the world average, and the highest rates of increase have been recorded since 1950. The diurnal range has been falling, as elsewhere. Rainfall seems to have increased in general, both on average and in the frequency of heavy rainfall events. In Australia, the largest increases have been on the east coast, with some decreases in the south-west and some inland areas. Heavy rainfall events are strongly related to *El Niño* fluctuations. The sea level has risen by around 2mm per year over the last 50 years.

For 2030, temperature increases of between 0.3 and 1.4°C are forecast for Australia and New Zealand, twice that for 2070. Precipitation for Australia seems uncertain, it may rise or fall, with the balance of probability thought to be for a fall. For New Zealand increases in the frequency of the prevailing westerly wind may increase precipitation in the west and reduce it in the east.[92]

More frequent high-rainfall events will cause flooding but are unlikely to improve the general water situation in Australia. In New Zealand, higher temperatures and higher atmospheric CO_2 concentration will generally increase yields. In forested areas there may be higher growth rates, while vulnerability to fire will certainly increase, so the net effect on forest cover is uncertain.

The most acute problems with sea-level rise are on outlying islands, though there are many urban and beach areas, and at least one national park area, that would be threatened. Increased flooding may result more from high-rainfall events. There is a wide range of estimates for changes in run-off rates, and higher variability is expected. Both Australia and New Zealand are expected to suffer losses of snow cover.

Europe

Europe (whose Eastern boundary is given by the Ural mountains and the Caspian Sea) has much intra-regional climate variation. It also has a rather varied experience of climate change over the past century: there has been some warming almost everywhere, though a few places have cooled a little (Greece and some parts of Eastern Europe); a belt extending from Spain through Central Europe to Russia has warmed a lot. Precipitation has generally risen in the north and fallen in the south.

Forecasts are not thought to be very reliable. An overall increase in winter temperatures is expected, by between 1.5° and 4°C according to the area and the model used. Summer temperatures may increase by up to 4.5°C, with a wider range of uncertainty than for the winter. The effect of aerosols is particularly important (and uncertain) in Central Europe – it might offset half the temperature increase due to CO_2. Precipitation overall also is expected to increase, particularly winter precipitation in the north. For the summer, some models show an increase, some a decrease.[93]

Agricultural productivity would probably be positively affected overall, though not necessarily everywhere; the interaction of changes in temperature and changes in average rainfall and its timing will be crucial. Growth rates in forests would probably increase (some predicted increases in growth rates already seem to be occurring), though the zones suitable for different types of trees may shift faster than the forest types can migrate, and some areas will become more vulnerable to fire. Peat producing areas will produce less peat and bogs may turn into fens. Many wetlands may change from being carbon sinks to sources. A number of areas of coast and coastal wetland are vulnerable.

Many glaciers will completely disappear and reductions in winter snowcover will affect many areas. The main impact of changes in precipitation is likely to be reduced summer run-off and increased winter run-off, even with higher annual precipitation – most of any increase in precipitation will come in the winter and evapo-transpiration[94] will increase. This will lead to increased problems with flooding, particularly since most floodplains are already "overpopulated"; coastal flooding due to storm surges will also increase with a higher sea level and a probable increase in frequency of storms.

5.2. *Sectoral impacts*

The degree of imprecision with respect to forecasts of even basic climate parameters is such that very little can be said with certainty about impacts on particular sectors, with the exception of impacts associated with rising sea levels. This section nevertheless considers what kinds of impacts one might expect, and where uncertainties may lie.

In all these areas there is a striking gap between the generally somewhat speculative, often qualitative rather than quantitative, predictions available and the kind of information needed to assign shadow prices (which may vary according to the different situations and preferences of different countries) to these impacts in order to refine the cost-benefit analysis of climate change and mitigation policies.

Agriculture

In most OECD countries (Australia may be an exception) warmer weather is likely to mean higher crop yields unless water availability is a constraint. The latter depends very much on the geographical and seasonal pattern of precipitation, as well as on temperature itself (which affects the rate of evapo-transpiration), forecasts for which are very uncertain. The rate of photosynthesis increases quite substantially in many crops (and trees) as the atmospheric CO_2 concentration increases. Drier and hotter conditions in some areas may nevertheless result in losses of agricultural land, but the northern retreat of tundra and permafrost zones in Canada, Northern Europe and Russia will presumably extend the northern margin of cultivated land. For the United States and Canada, and possibly for most of Japan, Europe, Korea and New Zealand, agricultural yields seem on average (with much local variation, no doubt) more likely than not to increase in response to climate change. Any increase in the frequency of extreme events (storms, floods, heatwaves, droughts) might work against this.

Fisheries

The most important impact on coastal fisheries is the potential loss of coastal wetlands, which are nurseries and nutrient suppliers, from sea-level rise. To a large extent, the amount of such wetlands that are lost will depend on choices about how to manage rising sea levels since for the most part they would naturally migrate inland as water levels rise. While species may migrate as water temperatures rise, there is no expectation of adverse effects on other fisheries. As the IPCC report notes, over-fishing and human intervention will probably have more influence on fisheries than climate change.

Transport

Only water transport seems likely to be directly affected by climate change. Changes in precipitation and its variability would affect river and canal transport; IPCC projections for North America and Europe suggest higher rainfall in spring and lower rainfall for the rest of the year, a pattern that would reduce the capacity of inland waterways. Sea-level rise will of course over a long period flood many existing port facilities. Ports currently suffering from closure or disruption by seasonal ice will generally benefit, however.

Water

Where higher temperatures are accompanied by lower precipitation, pressure on water supply, both quantity and quality, may increase. For most OECD countries this will be a purely domestic problem, though water rights can be a cross-border issue, as for example between the countries bordering on the Rhine in Europe. In some areas, tension between domestic, industrial and agricultural (see above) uses for water may increase. As for agriculture, changes in the geographical distribution of precipitation could have important effects, but little is known about such changes.

Energy

One can confidently predict relatively less demand for energy for space heating and rather more for refrigeration and air conditioning. Compared with trends generated by changes in tastes, wealth and technology, these shifts will be insignificant. Changes in precipitation patterns may affect hydroelectric generation, most likely to reduce potential if there is indeed a trend towards increased spring precipitation but less at other times of the year; a temporary, but possibly prolonged, benefit may arise from increased flows due to glaciers melting.

Insurance

The insurance industry exists because of uncertainty. If global warming does increase weather variability, particularly the incidence of storms and floods, the insurance industry may face increased demand.[95] As patterns change, however, the industry will have to pay particular attention to methods of calculating risks – historical averages on their own may be misleading.

Tourism

Tourism is most likely to be affected by the reduced amount of snow for winter sports. Significant rises in sea level will reduce available beaches, depending on the extent to which they can, or are allowed to, migrate landwards. Again it seems unlikely that these changes could be fast enough to be noticeable compared with shifts likely to come on the demand side.

Population movements

Since some coastal and estuarial locations would become submerged sea-level rise would necessitate some movement of population. Over the next

100 years the maximum expected rise of one metre would not result in much such pressure within OECD countries themselves, however. Potentially more important might be pressure for immigration from non-OECD countries suffering relatively more from climate change. There is little information from which to predict such pressures however; Africa contains a number of regions where populations may be most vulnerable to climate change but while IPCC projections imply warming throughout Africa there is no indication whether rainfall will increase or decrease. Also, IPCC notes that current agricultural yields are only about half of potential, implying that a fall in the latter is not necessarily associated with a fall in actual yields. As concerns population movements in response to sea-level rises, Bangladesh may be thought particularly vulnerable.

Health

The only likely systematic health impact from warming may be an increase in the geographical spread of vector-borne diseases as climates become more hospitable to mosquitoes and other disease-carrying insects. While one might expect warming to result in fewer winter-related deaths and more due to things like heat-stress, this is not clear since they are provoked more by variations in temperature than by average levels. It is also argued that there would be a larger increase in heat-related deaths than fall in winter deaths since the former are more closely related to temperature than the latter (Kupnick, 1998).

To judge by apparent preferences for vacations and, at least in North America, retirement, most people in temperate OECD countries prefer warmer weather. Unless this is more a relative than absolute preference, it may be supposed that warming in itself will improve well-being for many people, perhaps offsetting some of the higher health risks.

Biodiversity

The influence of global warming itself (as opposed to that of economic growth, industrialisation of agriculture, *etc.*) on biodiversity is not obvious (though it may be noted that biodiversity itself affects the ability of ecosystems to adapt to climate change). Where habitats are destroyed (as may be the case for some coral reefs and some forested areas) there will be losses but it is hard to say to what extent these may be offset by species migration and adaptation.

5.3. National *"impact assessments"*

To provide more detailed indications of likely weather patterns in particular regions or countries a higher degree of geographical resolution is needed than is feasible using the global climate models on which the previous paragraphs are based. One way of doing this is to use more detailed models restricted to one region, constraining them to reproduce the results of a global model around the boundary of the region in question. Using this approach, more detailed, but still preliminary, impact assessments have recently been produced for France and the United Kingdom.[96] It is noticeable that few other countries have produced such assessments – at least to the knowledge of the Secretariat.

The UK report – based on a "Business-as-Usual" scenario for emissions – suggests a general northward movement of temperature patterns, with the south of the country experiencing, by around 2050, temperatures similar to those of the present day south of France. But while more rain is expected overall, the already (relatively) dry south east could have much less summer rain; combined with higher temperatures and stronger winds this could dramatically change the landscape, increasing susceptibility to soil erosion and also damaging the foundations of many existing buildings. The UK report also draws attention to likely inundation of land along coasts and, perhaps more importantly, losses of low-lying high-grade agricultural land due to increased salinity. Not only agricultural land would be affected – most of the United Kingdom's petroleum refineries and nuclear power stations are vulnerable to rising sea levels. Many ecological systems would be threatened partly because of the speed of the projected changes and also because species migration is hampered. Greater damage from storms and other extreme events is also foreseen. At this stage, the report does not attempt to assess the overall economic costs of these projected developments.

The French report, whose climate simulation does not cover overseas departments and territories, also predicts some regional variation in impacts due largely to precipitation, while temperature zones move more or less systematically northwards (although differences between average temperatures in the south and the north may increase slightly). Agricultural output may increase overall, though crop patterns would have to adjust while being subjected to increased frequencies of storms and droughts. Temperature changes imply that forest types should migrate northwards, but some would be inhibited by lack of water; the Mediterranean zone would see an increased sensitivity to fire risk. Drier soils in summer, autumn and winter (higher soil moisture in spring is expected) will increase vulnerability to soil erosion as well as tending to deplete underground aquifers. Although tourism would be affected, particularly by a reduced Alpine skiing zone, the report does not suggest that the tourist industry would necessarily be damaged, though the relative attraction of different areas would no doubt change.

Both reports note that health is likely to be affected by climate change, though it is difficult to assess whether the overall impact would be beneficial or not. Illnesses associated with colder weather would obviously diminish, while other risks would increase. Higher temperatures will certainly favour increased vector-borne and infectious diseases in all countries, and illnesses associated with atmospheric pollution would also almost certainly increase (the production of ozone, which favours such illnesses, increases with higher temperatures).

5.4. *Adaptation, costs and catastrophe*

Climate change will certainly impose economic costs on some sectors, even though it may benefit others. Faced with prospective costs, it may be worth spending some resources either on trying to reduce emissions so as to stabilise the climate, or on adapting to the changed climate to reduce its impact. Chapters 3 and 4 have discussed the former, this section discusses adaptation.

Adaptation

The benefits of reducing emissions are public goods, with a strong presumption in favour not only of government intervention but of internationally co-ordinated government intervention. On the other hand, market forces may be better able to handle adaptation in many cases. There are grounds for public intervention, however, where externalities exist, and policy may need to facilitate market-led adaptation by ensuring that public policy does not unnecessarily distort price signals. Even where "market type" decisions are feasible in principle, in practice the necessary time horizon may be so long that public intervention may be necessary to some degree. In any case, to ensure that markets do react, public policy needs to provide a stable framework in which private long-term decisions will not be subject to excessive risk.

Agriculture and water supply (including for agriculture) are areas where some existing policies in some countries may well inhibit optimal adaptation. Adaptation in agriculture will include changing crop patterns as well as possibly ceasing to cultivate some land. The benefits of agricultural support policies that are tied to particular crops or which prevent land use changes will have to be weighed against the possible higher costs they impose on adaptation to climate change. Water supply charges will need to be allowed to evolve in a way that reflects changing overall or seasonal supply patterns. Adaptation in the energy market, tourism, insurance and many other sectors can probably safely be left to market mechanisms, though public policy may have some parallel adaptation to do, for example ensuring that prudential provision of the insurance sector is evolving appropriately. Moreover, it will be important to avoid public policies that attempt to support activities that are ultimately not sustainable but which may enjoy political leverage.

One area where public intervention is already prevalent is coastal management in the face of sea-level rise. Sea levels have been rising in many areas for some time, and some coasts suffer erosion even without such rises. Much experience has therefore accumulated in this area.

Three different types of strategy for coastal management in the face of sea-level rise can be distinguished: accommodation, planned retreat, and protection.[97] The first consists of making no attempt either to protect existing installations or to influence the rate of erosion, installations and facilities are adapted while possible and abandoned as they become unusable. The third consists of fully protecting existing installations through sea walls, flood barriers and so on, while the second, planned retreat, allows natural processes to continue, withdrawing or abandoning some installations in advance, protecting others where cost calculations suggest it is worthwhile, and possibly preventing development of some areas likely to be vulnerable or needed for adaptation in the future.

While planned retreat sounds the natural choice, its use in practice has been rare.[98] It requires a lot of information, including projections of sea level rise but also information on local interdependencies – for example between coastal wetlands which may diminish if dry land is protected, and fisheries which often depend on wetlands, as well as information on economic variables such as expected asset lives and necessary lead times for investment. Conflicts of interest are likely, and the legislative framework may need to be adapted.[99]

Coastal management also provides a precedent in terms of legislation to encourage forward planning. Most maritime countries have legislation that requires states or districts to draw up programmes to address rising sea levels.[100]

While projected temperature changes alone may not merit such comprehensive public-sector involvement as in coastal management, the example may be useful in the case of water resources, both where supplies may diminish substantially and where increased frequency of extreme events may have implications for floods, landslides, *etc*. Urban planning, construction standards and transport policy may all play a role in reducing future vulnerability to these changes. Where there is no obvious role for public-sector action, it may be worthwhile verifying that public policy does not inadvertently restrict the ability of individual agents to adapt.

An important aspect of adaptation is the speed at which climate change may occur. Warming of 2 to 3 degrees and sea-level rise of around 0.5 metre by the second half of the 21st century should not produce dramatic short-term adjustment problems for economic infrastructure. For example, the IPCC reports lead times of ten years for planning and construction of major facilities in Australia, *e.g.* dams and flood protection facilities; this seems a relatively short time, relative to the expected rate of change of the climate. Further, "the time frame for response to sea-level rise is, in principle, sufficient to allow for the development of suitable coastal policies...".[101] These conclusions may be cause for optimism, but while they suggest that mitigation and adaptation strategies are feasible, they do not in themselves imply that the costs will be low.

Costs

Since future emission paths are unknown, and since, even if they were known, the detailed consequences for climate and climate variability are unknown, calculating the costs of climate change is a very hypothetical exercise. In many areas, for example coastal management and water resources, fisheries, the pressure on resources of climate change may in any case be very small relative to existing and future pressures from population growth and economic development.

A certain number of attempts to make such calculations for countries or regions as a whole have been made, however, based on particular assumptions about emissions or climate change. The complexity of this task has ruled out model-based approaches up to now; rather, groups of experts collaborate on committee-type assessments[102] or secondary sources are used for costs in different sectors of the economy.[103] These exercises, which by their nature are of uncertain reliability, suggest that for most OECD countries the identified costs of climate change, up to the end of the 21st century, will be moderate. Thus, one estimate is that the costs of a doubling of CO_2 concentration for most OECD regions would be between 1.5 and 2 per cent of GNP, rather higher for Australasia, and that they could be between 5 and 10 per cent of GNP in Asia and Africa.[104] Other studies have results of a similar order of magnitude. Even the latter figures, when compared with possible productivity increases, do not seem enormous.[105] It is interesting that these estimates seem to be of the same broad order of magnitude as the estimated costs of stabilising concentration levels by the end of the 22nd century provided that the path towards stabilisation is not overly ambitious (see Chapter 4).

These estimates provide useful benchmarks, and mark an essential starting point in the search for quantification of the effects of climate change. But they are open to question. The figures are generally based on heroic assumptions about the unit cost of particular aspects of change, the various figures then being summed. Lacking more complete models of the links between climate and the economy (in both directions) it may not be possible to do much better.

One possible alternative is the so-called "Ricardo" approach, which looks at agriculture using cross-section data on farm performance across climate zones, attempting to correct for various influences such as the level of the capital stock and soil type, to isolate the influence of climate on agricultural productivity. Such studies for the United States, Brazil and India confirm suggestions that, if no account is taken of carbon fertilisation, the impact on developing country agriculture would generally be more severe than on that in OECD countries, and that adaptation can reduce estimates of damage by one-third or one-half; they also suggest that when the carbon fertilisation[106] effect is taken into account, the impact on developing country agriculture may be close to zero in aggregate (for particular areas, the impact may of course be severe).[107] Another interesting result is that the sensitivity of farm results to climate change does not depend on capital intensity.

It may be possible to adapt this methodology to other economic sectors, or to the economy as a whole, although agriculture may be unique in the wealth of the necessary highly geographically disaggregated data available. No such studies have been attempted so far.

Uncertainty

A potentially more serious objection to both of these approaches is that they ignore the impact of uncertainty on the assessment of costs. In particular, the probability distribution of impacts is almost certainly highly skewed: there are a number of unlikely but potentially catastrophic scenarios even in the medium term. Two that might concern most OECD countries are a collapse of part of the Antarctic ice shelf which could raise sea levels by several metres relatively quickly (compared with the usually assumed scenario of a rise of up to one metre over the next century), and severe disruption of the thermohaline circulation[108] in the north Atlantic and north Pacific, which could lead to a large and rapid cooling of climates on the north-east coasts of these oceans.

Although there is no easy approach to potential events whose probabilities are unknown and where even the consequences are hard to define, some account needs to be taken of them.[109] Even though the true probabilities are unknown, certain influences on them are known, so strategies that tend to reduce them can be undertaken. The Framework Convention on Climate Change is itself an example of action taken partly on these grounds, at least by a number of Parties for whom the apparently most likely consequences of climate change may be negative but not very serious.[110]

Influenced by the UNFCCC and by rising public awareness and concern, countries are moving towards more co-ordinated research into climate change impacts. For example, the United Kingdom established a Climate Impacts Programme; France set up the *Mission Interministérielle de l'Effet de Serre*. These programmes should in the future provide more complete pictures on which to base analysis.

Notes

1. Note that a short glossary is provided after Chapter 5 to remind readers of the meaning of some of the terms and acronyms that occur in this paper.

2. The IPCC was set up in 1988 as a joint body of the United Nations Environmental Programme and the World Meteorological Office. It monitors global climate developments and co-ordinates a programme of work aimed at improving scientific understanding of the mechanisms at work. Its assessment reports summarise the consensus view of the likely effect of current trends in emissions on future atmospheric concentration levels and the consequent effects on climate. The Second Assessment Report concluded, after a comprehensive survey of the scientific evidence and recognising that conclusive proof is not available, that there is a discernible influence of human activity on the climate, via the greenhouse effect; the report also investigated the properties of emission paths that would be necessary to stabilise atmospheric concentrations at various levels.

3. Most of the information in this section is based on the IPCC (1996).

4. The contribution to global warming of a gas is a combination of its concentration in the atmosphere and its relative effectiveness in trapping outgoing radiation. CO_2 currently provides over 60 per cent of the total anthropogenic greenhouse effect, methane about 20 per cent, nitrous oxide about 6 per cent and the chlorofluorocarbons about 10 per cent. See IPCC (1996) Table 4, pp 92-93, and the accompanying text for details and qualification of these estimates.

5. In the scientific literature, "anthropogenic" refers both to that which is directly the result of human activity (*e.g.* burning fossil fuels for heat or transport) and to the results of natural processes which have been modified by human activity (*e.g.* planting forests). This meaning is that intended throughout this paper.

6. The same relationships can be used in simple models to give an approximate idea of the effect on concentration of variations around the expected path of emissions, once a baseline (of historical and anticipated emissions and the resulting concentration levels) has been established (as in Chapter 4 below, for example).

7. See, for example, the assumed margins of error in the carbon "budget" estimate for the 1980s:

Sources and sinks of anthropogenic CO_2

Annual averages 1980-89, gigatonnes of carbon per year

	Amount	Confidence interval
Sources		
Fossil-fuel and cement production	5.5	(0.5)
Changes in land-use	1.6	(1.0)
Total	7.1	(1.1)
Sinks		
Atmosphere	3.3	(0.2)
Ocean	2.0	(0.8)
Northern hemisphere forest regrowth	0.5	(0.5)
Inferred sink	1.3	(1.5)

Source: IPCC, 1996, p. 17, Table 2.

8. They should really be known as Annex B countries, since it is Annex B to the Protocol that lists the emission targets. Of OECD countries, Korea, Mexico and Turkey are not listed in Annex B.
9. Belarus, Bulgaria, Croatia, Estonia, Latvia, Lithuania, Romania, Russia, Slovakia, Slovenia and Ukraine. Liechtenstein and Monaco are also included. Turkey, while currently listed in Annex 1, is not a Party to the Kyoto Protocol.
10. Conversion factors are based on the estimated impact of each gas in reducing outward radiation, relative to the impact of CO_2. Once converted to CO_2 equivalent, the measure itself is the weight of the carbon contained in the CO_2 equivalent; one tonne of CO_2 contains 0.27 (= 12/44, the ratio of the atomic weight of carbon to the molecular weight of CO_2) tonnes of carbon.
11. Iceland (+10%); Australia (+8%); Norway (+1%). See Annex B to the Protocol. An internal EU ("burden-sharing") plan to redistribute the quotas is allowed under the Protocol but has not yet been registered with the UNFCCC; under this EU agreement, country targets vary from –21% (Germany) to +27% (Portugal).
12. As recognised in the Protocol, a first-best policy to reduce emissions is to reduce subsidies to fossil-fuel consumption.
13. Found respectively in Articles 17, 6 and 12 of the Protocol.
14. The Annex 1 Expert Group, which meets under OECD/International Energy Agency auspices, plays an exploratory role in this process.
15. This paper refers to "permits" and "quotas", language not used in the Protocol but which is more transparent than that derived from the wording of the Protocol. Annex B to the Protocol defines the emissions target as a proportion of the total 1990 emissions, referred to as a QELRC or "Quantified Emission Limitation or Reduction Commitment"; from this can be calculated the emissions limit in terms of tonnes of carbon equivalent. Some papers refer to trading in "parts of Assigned Amounts", following language used in the Protocol itself, rather than in "permits".
16. Projects which increase take-up by sinks are also eligible.
17. 113 projects by June 1998. The hosts for these projects are either European transition countries or developing countries, the latter almost entirely in Latin America. See Joint Implementation Quarterly, September 1998.
18. The Protocol also provides for technological transfer and other assistance to non-Annex 1 countries.
19. See Puhl (1998).
20. Emission trading: "Any such trading shall be supplemental to domestic actions for the purpose of meeting [the Kyoto targets]".
    JI: "Any such project [must provide] a reduction in emissions, or an enhancement of removal by sinks, that is additional to any that would otherwise occur".
    "The acquisition of emission reduction units shall be supplemental to domestic action..."
21. The recent proposal made by the Council of the European Union is an example of such a suggestion (European Union – The Council, 1999). It set ceilings on both net acquisitions and net sales ("transfers") of emission rights under the three flexibility mechanisms.
22. Much less is known about the greenhouse properties of the non-CO_2 gases, especially the fluorine compounds, and the estimated warming potential may change as the state

of knowledge advances. However, for the purposes of the first commitment period, the conversion factors are fixed at the global warming potential outlined in the Kyoto Protocol. Revisions in the light of further information will affect the relative weights of the gases only as from the second commitment period.

23. Gielen and Kram (1998), Reilly *et al.* (1998), Brown *et al.* (1999).
24. See, *e.g.* OECD (1999*c*), for a wider discussion of these issues.
25. For example, OECD (1999*c*).
26. Entities exceeding their allowed SO_2 emissions pay a fine of $2 000 per ton, compared with a permit price of the order of around $100 per ton.
27. For non-compliance which is not deliberate, through miscalculation or mistaken policies for example, responses are likely to include assistance measures to help the country improve its policies, rather than punitive measures. These are found in the Ozone agreement, for example.
28. Among the current Annex 1 countries the most likely net sellers are some of the transition countries, notably Russia and Ukraine.
29. Frequently this is referred to as "seller" liability. "Issuer" may be more appropriate since the idea is that liability rests with the country who was originally allocated the permit.
30. This would lend increased importance to discussions on how and where to keep records of transactions in permits, how frequently they should be updated, and how public they should be.
31. Different combinations of Issuer and Buyer liability can also be imagined. If there were serious concern about trading increasing overall emissions, both Issuers and Buyers could be held fully liable.
32. It may also be noted that the options available could conceivably depend on whether trading was deemed by the WTO to be in a "commodity" or a financial instrument; in the former case permit trade would come under WTO rules and it might be difficult to use trade sanctions to encourage compliance. It would seem unreasonable, however, to classify emission permits as a commodity.
33. Such an arrangement might be known as the CO_2 Standard, or perhaps the Greenhouse Gas Board.
34. There is some variation, nevertheless, but errors are thought to be not more than ±2 to 5 per cent (Gielen and Kram, 1998).
35. Other models are WorldScan, MIT-EPA, POLES, RICE, CETA, SGM, MS-MRT, GTEM, MERGE, AIM, G-Cubed and the Oxford Model. For references see Glossary. Some of the models have an engineering based description of energy technology choices ("bottom-up approach") whereas others, including the OECD Secretariat's GREEN model, put more emphasis on general equilibrium properties while treating energy technology in a more aggregate fashion ("top-down approach").
36. Although the current US negotiating position implies that the Kyoto protocol would not be ratified in the absence of the flexibility mechanisms.
37. For the sake of simplicity, and reflecting the structure of some of the models involved, the first commitment period is taken to correspond to its mid-point, *i.e.* 2010.

38. Among the successor states to the Soviet Union only the Russian Federation and Ukraine have signed the Kyoto Protocol. These countries are by far the dominant emitters in the GREEN CIS region.
39. For reference, the international oil price expressed in terms of carbon equivalent amounted to $150 per ton in 1995.
40. GREEN results depart from this view with higher marginal abatement cost in the United States than in the European Union and Japan (Table 4). This reflects the GREEN BaU scenario, which implies higher emission growth in the United States (2 per cent on average up to 2010) than in the European Union (1.2 per cent) and Japan (1.6 per cent). In consequence, the cut relative to baseline emissions is relatively much larger in the United States than in other OECD countries. The different emission trends in the BaU are accounted for mainly by demographic trends. The BaU scenario is obviously subject to uncertainty, but the growth differential of emissions between the United States and the European Union as projected in GREEN (0.8 percentage point on average during the period 1995 to 2010) is roughly in line with the corresponding trends observed during the early 1990s (1.2 percentage points). If these trends were to persist over the near future, they would imply a much larger abatement effort in the United States than in the European Union in order to meet the Kyoto commitments.
41. Some models listed in Table 4 incorporate adjustment costs (the Oxford Model is an example) and therefore suggest higher economic costs, up to around 2 per cent annually.
42. Most models reported in Table 4 show no carbon taxes in the CIS. The carbon increase of 4 per cent reported by the GREEN model is the result of feedback effects from limitations in the other Annex 1 countries.
43. Exceptions are the macroeconomic Oxford Model and the model G-Cubed which incorporates nominal rigidities of wages (McKibbin *et al.*, 1999).
44. In GREEN, the production has a putty/semi-putty structure and a distinction is made between new and vintage (installed) capital. The degree of substitution associated with vintage capital is lower than for new capital. In addition, vintage capital is only partially mobile across sectors.
45. This proportion is dependent on the degree of disaggregation of the model, with GREEN being fairly aggregated in structure and therefore tending to show rather low estimates of labour reallocation. Moreover, the necessary reallocation depends on the initial composition of employment and output across sectors, as well as the extent to which relative prices change, which may differ between countries and regions.
46. For instance, OECD (1999*a*); Turner *et al.* (1993).
47. This corresponds to the price of the "fuel and power" consumer aggregate in GREEN. The lower price effect in Japan reflects the large non-energy component in the average consumer price for "fuel and power".
48. Alternatively, recirculation of revenue could take place via reductions in distortive taxes, thereby reducing the overall economic costs of action to reduce CO_2 emissions.
49. This includes the leakages that would occur in the CIS due to the existence of "hot air", although leakages within Annex 1 countries have ultimately no impact on concentration due to the "banking" provision.
50. This categorisation has been used by, *e.g.* Frankel (1998).

51. Studies generally do not distinguish trading of emissions allowances from JI projects: they treat both instruments as if they had the same impact.
52. Part of the cost reduction in the scenario simulated by Manne and Richels comes from the implementation of the Clean Development Mechanism.
53. In GREEN, permit buyers tend to suffer adverse terms-of-trade effects for three reasons:

    1) GREEN operates under the assumption that capital flows between countries/regions remain constant in real terms. As countries buy permits from abroad, they have to reduce imports and/or to increase exports of other commodities in order to satisfy the balance-of-payment constraint. As many models, GREEN contains the assumption that non-energy goods are differentiated by origin (the so-called Armington specification). Therefore, countries behave as price setters on trade markets and restoring the external balance requires depreciation of the real exchange rate, implying a terms-of-trade loss.
    2) As countries buy permits, they increase their oil imports which yields the same kind of adjustment as in 1).
    3) As emission trading among Annex I countries shifts the abatements from oil to coal, the international oil price increases.

    The same three mechanisms operating in the opposite direction explain why the CIS benefits from an appreciation of its terms of trade.
54. See European Union, the Council (1999).
55. See Bollen *et al.* (1998); Tulpulé *et al.* (1998); Me Kibbin *et al.* (1998); Kainuma *et al.* (1998); US Administration (1998).
56. A preliminary analysis by the International Energy Agency (OECD, 1999*b*) suggests that the impact of the proposal on the EU will be to cap net purchases of emission rights to about 42 per cent of the projected difference between BaU emissions and the Kyoto target for 2008-12. For other industrialised countries the cap would amount to about 34 per cent whereas Russia would be prevented from selling more than 30 per cent of its projected "hot air".
57. This large cost increase is explained by the fact that the caps scenario is compared to a situation with world emission trading – *i.e.* assuming that the CDM operates like fully flexible emission trading. However, the corresponding GDP loss remains rather modest (45 billions of 1990 dollars in 2010).
58. Assuming that the quota rent is allocated to permit sellers (see below).
59. The broad orders of magnitude involved in this simulation are not out of line with the OECD (1999*b*) interpretation and quantification of the size of caps on permit purchases implied by the EU proposal (European Union, The Council, 1999), given the assumption that the CDM is not used.
60. Ceilings on the use of the CDM may have a larger impact, depending on the extent to which the CDM is exploited in practice. In the case where the CDM is used to the full of its potential, OECD countries would rely much more on the CDM than on the two other mechanisms in order to comply with the Kyoto targets. Assuming full trading at the world level, the proportion of emission purchases by the OECD countries would amount to around 90 per cent of their Kyoto commitments expressed as reductions relative to the baseline in 2010.
61. Assuming that the CIS receives the associated rent (see below).

62. According with GREEN simulations, the EU proposal would imply that CIS sales could not exceed 47 million tons of carbon by 2010 (roughly a third of the total amounts of "hot air"). Assuming unrestricted emissions transfers among Annex 1 countries, the CIS would sell around 400 million tons of carbon in 2010. Thus, the EU proposal implies a reduction of Annex 1 emission transfers by a factor of ten. Simulation results indicate that it would almost completely obliterate the efficiency gains expected from the use of the flexibility mechanisms.
63. Alternative assumptions on how the rent accruing from trade restrictions is distributed matter for the impacts on real incomes (Manne and Richels, 1998).
64. For instance, according to the SGM model, full Annex 1 trading together with the CDM would cut the cost to the United States by 66 per cent. In comparison, trading among Annex 1 countries alone would reduce the US cost by 57 per cent.
65. Entry into force requires ratification by at least 55 countries, including Annex 1 countries whose emissions total at least 55 per cent of total Annex 1 CO_2 emissions in 1990.
66. The current US position is that ratification is contingent on there being no serious restrictions on emission trading, and on participation by some major developing countries. Judging by the discussion at COP4, the latter condition seems unlikely to be met in the near future.
67. The GREEN model takes into account that new backstop technologies may emerge starting after 2010. The model embodies rigidities in capital turnover (through its putty-clay production functions), but it does not include forward-looking behaviour and the fact that the model solves in five-year steps reduces the choice of potential time-paths that can be considered.
68. In particular, this reflects the assumption that no backstop energy sources are available by 2010.
69. These assessments have been made using a 3 per cent discount rate.
70. Some gases have a much larger warming potential than others. For instance, the warming potential of SF_6 is 24 000 times higher than CO_2.
71. Only the MIT-EPPA model (Reilly *et al.*, 1998) and the GTEM model (Brown *et al.* 1999) incorporate some non-CO_2 GHGs.
72. Corresponding to the end of the 18th century.
73. This study was made prior to the signing of the Kyoto Protocol. It analyses the implications of the proposals submitted by the Annex 1 countries to the COP3 meeting.
74. Three different models have been used in order to examine the robustness of the results. This text is based on a model which is derived from Wigley (1997) and is linear in that the carbon remaining in the atmosphere is expressed as a linear function (*i.e.* a fraction) of the initial emission (see Annex 2).
75. It may be argued that there are some signatories of the Kyoto Protocol - Canada and the Russian Federation are possible examples - which, according to the rationality of the game theory model, should never have joined an agreement to reduce emissions because they could stand to gain from a warming of the climate.
76. Not as such; ancillary damage due to local or other pollution effects has of course been recognised in various ways.
77. This notion of convergence through time in *per capita* emissions was discussed at COP4, see IEA (1998).

78. Assuming that countries that are not subject to an emission constraint, are allowed to sell emission permits based on their BaU emissions. This aims at approximating the full potential of the CDM.
79. This roughly approximates the costs of a stable grand coalition. However, it may be a lower bound estimate as it does not take into account the existence of possible carbon leakages, which would imply a need for larger inducements to avoid free-riding. Local, ancillary benefits, however, may point in the opposite direction, *i.e.* to this being an upper-bound cost estimate.
80. Assuming a discount rate of 3 per cent per year.
81. In the simulations, costs differ slightly across rules due to the dynamic of capital accumulation which differs across countries/regions.
82. Calculated with a 3 per cent discount rate.
83. It may happen that the marginal cost gain obtained by buying permits is offset by the terms-of-trade deterioration induced by buying large amounts of permits. See Annex 3.
84. Account being taken of the proceeds from permits sales.
85. These are lower bound estimates as they do not incorporate potential carbon leakages. The assumption made here is that non-Annex 1 countries receive funds corresponding to their loss relative to the BaU real income level. However, this may not entirely offset the incentive for free-riding since non-participating countries may gain by staying outside a coalition due to spillover effects through trade and energy prices. In principle, the compensation payment to guarantee the stability of the grand coalition should be calculated as the difference between the real income level if the country stays outside the coalition (which may be higher than the BaU real income) and the corresponding real income if the country participate to the grand coalition. According to GREEN, however, carbon leakages are not very important, especially when the acting coalition involves all countries except one (see the discussion on page 35 and Box 4). However, local ancillary benefits may reduce the need for financial transfers to induce participation.
86. These gains arise because some non-Annex 1 countries receive compensation for any losses they incur, whereas other non-Annex 1 countries that are not capped can improve their real income by selling permits based on their baseline emissions (the CDM mechanism).
87. This reflects the heterogeneity of the non-Annex 1 group of countries: in particular, the "grandfathering" rule shifts the burden of abatement to low cost developing countries (China and India) while the two other rules require higher commitments in oil-consuming semi-industrialised countries (which have higher marginal abatement costs).
88. Africa; Middle East and Arid Asia; Tropical Asia; Small Island States; The Arctic and Antarctic.
89. For Japan, Korea and Mexico, little specific information can be deduced from the regional reports since they are geographically small parts of the relevant regions (Latin America and Temperate Asia).
90. In the sense that this amount of land is less than 0.5 metres above current high-tide levels.
91. The report also notes that assets located in the respective areas are worth $450 billion and $908 billion, but the basis of this valuation is not clear. More meaningful might be

the estimated $80 billion necessary to modify relevant flood protection schemes to protect the expanded area.

92. It is interesting to note that the previous set of IPCC forecasts for New Zealand foresaw the opposite – a reduction in the frequency of westerly winds and thus a reduced imbalance between precipitation in the east and the west. The change is due to the availability of more detailed models of ocean circulation.
93. Climate change in Europe would be much more dramatic and could be sudden if the north Atlantic thermohaline ocean circulation (which brings the Gulf Stream to the coasts of north-western Europe) were to cease or diminish substantially. The probability of this is thought to be very low for the degree of climate change thought likely over the next century or so.
94. Evaporation of water directly from the ground and via plant activity.
95. IPCC (1998) reports that the cost of insured catastrophes in Australia is at least 12 per cent of the costs of non-life premiums, compared with 2.5 per cent in the rest of the world– the result of relatively high frequency of climatic and other natural hazards. This is despite the unavailability of private flood insurance for dwellings and small business. Some of this difference could of course be due to attitudes to risk and the structure of the insurance industry, rather than due to different or higher risks.
96. Mission Interministérielle de l'Effet de Serre, 1998, and Department of the Environment, Transport and the Regions, 1998, respectively.
97. See, for example, IPCC (1998) pp. 173-174 and pp. 302-303.
98. IPCC (1998) mentions experiments in eastern England where sea walls have been deliberately breached, turning previously protected farmland into coastal wetlands; this is intended to relieve pressure on the sea defences, to provide flood alleviation benefits and to help restore natural balance in estuaries. See IPCC (1998) p. 174.
99. Some measures by US states to restrict the rights of owners of coastal land to prevent the inland migration of beaches and wetlands have had to be modified after court challenges. See IPCC (1998) p. 303.
100. For example, The US Coastal Zone Management Act, the New Zealand Resource Management Act.
101. IPCC (1996), p. 133.
102. For example, Mosley, M.P., ed., Climate Change in New Zealand, 1990, quoted in IPCC, 1998.
103. Fankhauser (1995) (and Fankhauser and Tol, 1997; and Fankhauser *et al.*, 1998) takes into account the costs of coastal defences, loss of dry land and of wetlands, loss of ecosystems, the effect on agriculture and forestry, the energy and water sectors, the cost of mortality changes and those of air pollution, migration and hurricane damage. In principle, the estimates assume that optimal adaptation responses are forthcoming. Some indication of the importance of this assumption can be found in studies of the impact on agriculture which show that "welfare losses" are frequently reduced by half, relative to a no-behaviour-change baseline, when adaptation is assumed.
104. Fankhauser *et al.* (1998), quoting earlier work by the same authors.
105. The IPCC regional reports do not systematically give similar estimates of costs, although some quote the figures in Fankhauser and Tol (1997). The one country-specific study reported is of Finland, which may benefit by about 1 per cent of GDP (Kuoppomaeki, 1996, quoted in IPCC, 1998).

106. See Glossary.
107. See Mendelsohn *et al.* (1994), Dinar *et al.* (1998). A study covering a number of European and Mediterranean countries is currently under way.
108. See Glossary.
109. One way of looking at the long-term impacts of climate change policies is the so-called Tolerable Windows Approach. This is based on an inverse modelling concept where one decides on what represents an unacceptable future risk, as a function of future greenhouse gas concentration levels and then works backwards to look at feasible paths for emissions paths. The costs of policies to establish such paths can then be assessed against the degree of risk that is thereby avoided. This does not avoid the problems caused by our considerable ignorance about future climate change, but has the advantage of focusing attention on reducing risk (or indicators of risk) and the costs of so doing. See Toth *et al.* (1998), one conclusion which is that while different short-term emission paths may not make much difference to the costs of eventual climate change, they can have an important impact if the rate of change of climate matters.
110. There are many other reasons, including altruism *vis-à-vis* countries who expect to suffer more.

Bibliography

Babiker, M. and H.D. Jacoby (1999),
"Developing country effects of Kyoto-type emissions restrictions", mimeo., Joint Program on the Science and Policy of Global Change, Massachusetts Institute of Technology.

Barret, S. (1992),
"Conventions on climate change: economic aspects of negotiations", OECD, Paris.

Barret, S. (1994),
"Self-enforcing international environmental agreements", *Oxford Economic Papers*, No. 46, pp. 878-94.

Barret, S. (1995),
"Trade restrictions in international environmental agreements", London Business School.

Barret, S. (1997),
"Heterogeneous international environmental agreements", in C. Carraro (ed.), *International Environmental Agreements: Strategic Policy Issues*, E. Elgar, Cheltenham.

Bayar, A.H. (1998),
Can Europe Reduce Unemployment Through Environmental Taxes? A General Equilibrium Analysis, draft presented at the Twelfth International Conference on Input-Output Techniques, International Input-Output Association, New York, May.

Bernstein, P.M., W.D. Montgomery and T.F. Rutherford (1999),
"Global impacts of the Kyoto Agreement: results from the MS-MRT Model", paper presented at the IPCC Working Group III Expert Meeting, May, The Hague.

Bollen, J., A. Gielen and H. Timmer (1998),
"Compliance with the Kyoto Protocol", in *Economic Modelling of Climate Change*, OECD Workshop Report, 17-18 September.

Bollen, J., T. Manders and H. Timmer (1999),
"Kyoto and carbon leakage: simulations with WorldScan", mimeo., paper presented at the IPCC Working Group III Expert Meeting, May, The Hague.

Botteon, M. and C. Carraro (1998),
"Environmental coalitions with heterogeneous countries: burden-sharing and carbon leakages", *Nota di Lavoro* 24.98, March.

Brown, S.D., D. Kennedy, D. Polidano, K. Woffenden, K. Jakeman, G. Graham, F. Jotzo and B.S. Fisher (1999),
"Economic impacts of the Kyoto Protocol: accounting for the three major greenhouse gases", Australian Bureau of Agricultural and Resource Economics (ABARE), Research Report 99.6, Canberra.

Burniaux, J.M. (1998),
"How important is market power in achieving Kyoto: an assessment based on the GREEN model", in *Economic Modelling of Climate Change*, OECD Workshop Report, 17-18 September.

Burniaux, J.M., J. Martin, G. Nicoletti and J. Oliveira-Martins (1992),
"GREEN: a multi-sector multi-region dynamic general equilibrium model for quantifying the costs of curbing CO_2 emissions: a technical manual", OECD *Economics Department Working Papers*, No. 116, Paris.

Burniaux, J.M. and C. Complainville (1999),
"GREEN: a global model for quantifying the costs of policies to curb CO_2 emissions: an updated version", forthcoming as an OECD *Economics Department Working Paper*.

Burniaux, J.M. and J. Oliveira-Martins (1999),
"Carbon emission leakages: an analytical general equilibrium view", draft presented at the 2nd Annual Conference on Global Economic Analysis, 20-22 June, Copenhagen.

Cao M., F.I. Woodward (1998),
"Dynamic responses of terrestrial ecosystem carbon cycling to global climate change", Nature, 21 May, Vol. 393, pp. 249-252.

Capros, P. (1998),
"Economic and energy system implications of European CO_2 mitigation strategy: synthesis of results from model based analysis", in *Economic Modelling of Climate Change*, OECD Workshop Report, 17-18 September.

Capros, P. and E. Kokkolakis (1996),
"Energy efficiency and conversion decentralisation: evidence from the PRIMES model", International Workshop jointly organised by the European Commission, the International Energy Agency and the OECD on Instruments for Environmental Improvement with Structural and Technological Change in the Electricity Sector, 9-10 September, Belgium.

Capros, P., T. Georgakopoulos, D. Van Regermorter, S. Proost, T. Schmidt and K. Conrad (1997),
"The GEM3-E3 model for the European Union", *Journal of Economic and Financial Modelling*, Vol. 4, Nos. 2 and 3, special double issue, pp. 51-160.

Carraro, C. (1998*a*),
"The structure of International Environmental Agreements", *Nota di Lavoro* 12.98, January.

Carraro, C. (1998*b*),
"Beyond Kyoto: a game theoretic perspective", in *Economic Modelling of Climate Change*, OECD Workshop Report, 17-18 September.

Carraro, C. and D. Siniscalco (1992),
"Transfers and commitments in international environmental negotiations", forthcoming in KG. Mäler (ed.), *International Environmental Problems: an Economic Perspective*, Kluwer Academic Publishers: Dordrecht.

Carraro, C. and D. Siniscalco (1993),
"Strategies for the international protection of the environment", *Journal of Public Economics*, No. 52, pp. 309-28.

Carraro, C. and D. Siniscalco (1995),
"Policy coordination for sustainability: commitments, transfers and linked negotiations", in I. Goldin and A. Winters (eds.), *The Economics of Sustainable Development*, Cambridge University Press, Cambridge.

Carraro, C. and F. Moriconi (1998),
"Endogenous formation of environmental coalitions", draft, January.

Criqui, P. *et al.* (1996),
POLES 2.2, European Commission, DGXII, EUR 17538 EN.

Department of the Environment, Transport and the Regions (1998),
"Climate change impacts in the UK", London.

Dinar, A., R. Mendelsohn, R. Evenson, J. Parikh, A. Sanghi, K. Kumasr, J. McKinsey and S. Lonergan (1998),
"Measuring the impact of climate change on Indian agriculture", *World Bank Technical Paper* No. 402.

Enting, I.G., T.M.L. Wigley and M. Heimann (1994),
"Future emissions and concentrations of carbon dioxide: key ocean/atmosphere/land analysis", CSIRO division of atmospheric research technical paper, CSIRO, Australia.

European Union – The Council (1999),
Community Strategy on Climate Change: Council Conclusions, 8346/99, DGI, May.

Fankhauser, S. (1995),
"Valuing climate change. The economics of the greenhouse", London, *Earthscan*.

Fankhauser, S. and R. Tol. (1997),
"The social costs of climate change: the IPCC Second Assessment Report and beyond", *Mitigation and Adaptation Strategies for Global Change*, Vol. 1, No. 4, pp. 385-403.

Fankhauser, S., R. Tol and D. Pearce (1998),
"Extensions and alternatives to climate change impact valuation: on the critique of IPCC Working Group III's impact estimates", *Environment and Development Economics*, Vol. 3, pp. 59-81.

Frankel, J.A. (1998),
"What kind of research on climate change economics would be of greatest use to policy-makers?" in *Economic Modelling of Climate Change*, OECD Workshop, 17-18 September.

Geurts, B., A. Gielen, R. Nahuis, P. Tang and H. Timmer (1995),
"WorldScan: an economic world model for scenario analysis", *Review of WorldScan*, CPB, The Hague.

Geurts, B., A. Gielen, R. Nahuis, P. Tang and H. Timmer (1997),
WorldScan: project report to the National Research Program on Global Air Pollution and Climate Change, Project No. XXX, Bilthoven, the Netherlands.

Gielen, A. and C. Koopmans (1998),
"The economic consequences of Kyoto", CPB *report*, Vol. 98, No. 1, pp. 30-33.

Gielen, D. and T. Kram (1998),
"The role of non-CO_2 greenhouse gas in meeting Kyoto Targets", in *Economic Modelling of Climate Change*, OECD Workshop Report, 17-18 September.

Grubb, M., A. Michaelova, B. Swift, T. Tietenberg, Z. Zhang and F.T. Joshua (1998),
Greenhouse Gas Emissions Trading: Defining the Principles, Modalities, Rules and Guidelines for Verification, Reporting and Accountability, UNCTAD.

Ha-Duong, M., M.J. Grubb and J.C. Hourcade (1997),
"Influence of socioeconomic inertia and uncertainty on optimal CO_2 emission abatement", *Nature*, Vol. 390, pp. 270-73.

Hahn, R.W. (1984),
"Market power and transferable property rights", *The Quarterly Journal of Economics*, Vol. 99, pp. 753-765.

Hertel, T.W. (1997),
Global Trade Analysis: Modeling and Applications, Cambridge University Press.

Hoel, M. (1991),
"Global environmental problems: the effects of unilateral actions taken by one country", *Journal of Environmental Economics and Management*, Vol. 20, No. 1, pp. 55-70.

IEA (1998),
"Standing Group on Long-Term Co-operation, COP4 Outcome", IEA/SLT(98)49, OECD, Paris, 23 November.

IPCC (1996),
"Climate change 1995: the science of climate change. Contribution of Working Group 1 to the Second Assessment Report of the IPCC", J.J. Houghton, L.G. Meiro Filho, B.A. Callander, N. Harris, A. Kattenberg and K. Maskell (eds.), *Cambridge University Press*, Cambridge.

IPCC (1998),
"The regional impacts of climate change: an assessment of vulnerability". A Special Report of IPCC Working Group II.

Jain, A.K., H.S. Kheshgi, M.I. Hoffert and D.J. Wuebbles (1995),
"Distribution of radiocarbon as a test of global carbon cycle models", *Global Biogeochem Cycles*, No. 9, pp. 153-66.

Joint Implementation Quarterly (1998),
Joint Implementation Network, Netherlands, September.

Kainuma, M., Y. Matsuoka and T. Morita (1998),
"Recent analysis of emission scenarios based on the AIM model", handout of the presentation made at the Joint Meeting of the International Energy Workshop and the Energy Modeling Forum, Stanford University, 17-19 June.

Katsoulacos, Y. (1997),
"R&D spillovers, R&D cooperation, innovation and international environmental agreements", in C. Carraro, (ed.) *International Environmental Agreements: Strategic Policy Issues*, E. Elgar, Cheltenham.

Kolstad, C., M. Light and T.F. Rutherford (1999),
Coal Markets and the Kyoto Protocol, mimeo., April.

Kuoppomaeki, P. (1996),
"Impact of climate change from a small Nordic country perspective", ETLA – The Research Institute of the Finnish Economy, Series B119, Helsinki.

Kupnick, A. (1998),
"Climate change health risks and economics", *Resources for the Future*, Washington.

Lee, H., J. Oliveira-Martins and D. van der Mensbrugghe (1994),
"The OECD GREEN Model: an updated overview", OECD *Development Centre Technical Paper*, No. 97, August, Paris.

Manne, A., R. Mendelsohn and R.G. Richels (1995),
"MERGE – a Model for Evaluating Regional and Global Effects of GHG reduction policies", *Energy Policy*, Vol. 23, No. 1, pp. 17-34.

Manne, A. and R.G. Richels (1998),
"The Kyoto Protocol: a cost-effective strategy for meeting environmental objectives?" in *Economic Modelling of Climate Change*, OECD Workshop Report, 17-18 September.

McKibbin, W.J., P.J. Wilcoxen (1995),
"The theoretical and empirical structure of the G-Cubed Model", Brooking Discussion Paper in International Economics #118, The Brookings Institution, Washington DC.

McKibbin, W., R. Shackleton and P.J. Wilcoxen (1997),
"The international trade and financial impacts of carbon emissions reductions", draft, August.

McKibbin, W., R. Shackleton and P.J. Wilcoxen (1998),
"The potential effects of international carbon emissions permit trading under the Kyoto Protocol", in *Economic Modelling of Climate Change*, OECD Workshop Report, 17-18 September.

McKibbin, W., M.T. Ross, R. Shackleton and P.J. Wilcoxen (1999),
"Emissions trading, capital flows and the Kyoto Protocol", paper presented at the IPCC Working Group III Expert Meeting, May, The Hague.

Mendelsohn, R., W.D. Nordhaus and D. Shaw (1994),
"The impact of global warming on agriculture: a Ricardian analysis", *American Economic Review*, Vol. 84, pp. 753-71.

Michaelis L. (1996),
"Reforming coal and electricity subsidies", Policies and Measures for Common Action, Working Paper No. 2, Annex 1 Expert Group on the UN FCCC, July 1996, OECD/IEA, Paris.

Mission Interministérielle de l'Effet de Serre (1998),
"Impact potentiel du changement climatique en France au XXIe siècle", Paris.

Mosley, M.P (1990) (ed.),
"Climate change in New Zealand", New Zealand Ministry of Environment.

Mullins, F. (1998),
"International emissions trading under the Kyoto Protocol", Annex 1 Expert Group on the UNFCC, draft Information Paper, Environment Directorate, OECD, 2 September.

Nilsson, S. and W. Schopfhauser (1995),
"The Carbon-sequestration potential of a global afforestation program", *Climatic Change*, No. 30, pp. 267-93.

OECD (1995),
Climate Change, Economic Instruments and Income Distribution, Paris.

OECD (1996),
OECD *Employment Outlook*, Paris, July.

OECD (1997*a*),
Reforming Energy and Transport Subsidies: Environmental and Economic Implications, Paris.

OECD (1997*b*),
"Environmental implications of energy and transport subsidies, Volume 2, Supports to the coal industry and the electricity sector", OCDE/GD(97)155, Paris.

OECD (1997*c*),
"Environmental implications of energy and transport subsidies, Volume 1, Scoping Study, Greenhouse gas impacts of Russian energy subsidies, Climate change impacts of subsidies to the energy sector in the USA", OCDE/GD(97)154, Paris.

OECD (1998*a*),
"Economic Modelling of Climate Change", OECD Workshop Report, Held at OECD Headquarters, 17-18 September.

OECD (1998*b*),
"Ensuring compliance with a global climate change agreement", ENV/EPOC(98)5/REV/1, Paris.

OECD (1999*a*),
EMU: *Facts, Challenges and Policies*, Paris.

OECD (1999*b*),
"A preliminary analysis of the EU proposals on the Kyoto mechanisms", note prepared by R. Baron, M. Bosi, A. Lanza and J. Pershing, International Energy Agency, Paris, May.

OECD (1999*c*),
"International emissions trading under the Kyoto Protocol", OECD Information Paper, ENV/EPOC(99)18/FINAL.

OECD (1999*d*),
"Ensuring compliance with a global climate change agreement", OECD Information Paper, ENV/EPOC(98)5/REV1.

OECD (1999*e*),
"Trade measures in multilateral environmental agreements: synthesis report of three case studies", Joint Session of Trade and Environment Experts, COM/ENV/TD998)127.

Puhl, I. (1998),
"Status of research on Project Baselines under the UNFCCC and the Kyoto Protocol", OECD and IEA Information Paper.

Reilly, J., R.G. Prinn, J. Harnisch, J. Fitzmaurice, H.D. Jacoby, D. Kicklighter, P.H. Stone, A.P. Sokolov and C. Wang (1998),
"Multi-gas assessment of the Kyoto Protocol", draft.

Siegenthaler, U. and F. Joos (1992),
"Use of a simple model for studying oceanic tracer distributions and the global carbon cycle", *Tellus*, No. 44, pp. 186-207.

Toth, F.L., G. Petschel-Held and T. Bruckner (1998),
"Kyoto and the long-term climate stabilisation, in *Economic Modelling of Climate Change*, OECD Workshop Report, 17-18 September.

Tulpulé, V., S. Brown, J. Lim, C. Polidano, H. Pant and B.S. Fisher (1998),
"An economic assessment of the Kyoto Protocol using the Global Trade and Environment Model", in *Economic Modelling of Climate Change*, OECD Workshop Report, 17-18 September.

Turner, D., P. Richardson and S. Rauffet (1993),
"The role of real and nominal rigidities in macroeconomic adjustments: a comparative study of the G3 economies", OECD *Economic Studies*, No. 21, Winter.

Turner, D., C. Giorno, A. De Serres, A. Vourc'h and P. Richardson (1998),
"The macroeconomic implications of ageing in a global context", OECD *Economics Department Working Paper*, No. 193.

US Administration (1998),
"The Kyoto Protocol and the President's policies to address climate change", draft, July.

Weyant, J.P. and J.N. Hill (1999),
"Kyoto special issue, Introduction and Overview", forthcoming in the *Energy Journal*, International Association for Energy Economics.

Wigley, T.M.L. (1993),
"Balancing the global carbon budget. Implications for projections of future carbon dioxide concentration changes", *Tellus*, Vol. 45, pp. 45-48.

Wigley, T.M.L., (1997),
"Implications of recent CO_2 emission-limitation proposals for stabilization of atmospheric concentrations", *Nature*, Vol. 390, 20 November.

Wigley, T.M.L., R. Richels and J.A. Edmonds (1996),
"Economic and environmental choices in the stabilization of atmospheric CO_2 concentrations", *Nature*, Vol. 379, 18 January.

Wigley, T.M.L., A.K. Jain , F. Joos, B.S. Nyenzi and P.R. Shukla (1997),
"Implications of proposed CO_2 emissions limitations", Intergovernmental Panel on Climate Change.

World Bank (1997),
"Expanding the measure of wealth: indicators of environmentally sustainable development", *Study and Monography Series*, No. 7, pp. 40-65.

Glossary and Abbreviations

Assigned Amount

Permitted emissions for the period 2008-12 in tonnes of carbon equivalent.

Carbon fertilisation

The rate at which photosynthesis proceeds, and hence the rate of plant growth, in most plants is, in laboratory conditions, an increasing function of the amount of CO_2 in the air, at least up to a certain point and provided growth is not constrained by lack of nutrients or too high temperatures. The extent to which this would be true in practice is uncertain, especially for mature trees.

Commitment Period

Period for which actual emissions are compared with the Assigned Amount to assess compliance with emission limits. The first is 2008-12, the expectation is that the next would be 2013-17.

COP

Conference of the Parties (to the UNFCCC).

Evapo-transpiration

Water vapour gets into the atmosphere from the ground either through direct evaporation or through being taken up by plants and subsequently being emitted through their leaves; the combination of these processes is evapo-transpiration.

IPCC

Intergovernmental Panel on Climate Change. Established in 1988 as a joint body of the United Nations Environmental Programme and the World Meteorological Office.

Models

The paper refers to a number of models used in assessing aspects of climate change and policies:

AIM (Asian-Pacific Integrated Model): integrated assessment model including a recursive general equilibrium model of the world economy, a warming forecasting model and a warming impact model. This model has been developed by the National Institute for Environmental Studies and the Faculty of Engineering of the Kyoto University.

CETA: Carbon Emissions Trajectory Assessment Model developed by the Electric Power Research Institute.

EMF-16: the Energy Modelling Forum (EMF) was established in 1976 at Stanford University to provide a structural framework within which energy experts, analysts, and policy makers could meet to improve their understanding of critical energy problems. The forthcoming sixteenth EMF study (EMF-16) will focus on "Integrated Assessment of Climate Change: Post-Kyoto analysis".

EPPA: Emissions Projections and Policy Analysis Model, a general equilibrium model developed by the Massachusetts Institute of Technology.

G-Cubed: a multi-region, multi-sector intertemporal general equilibrium model initially developed by W. McKibbin and P.J. Wilcoxen (McKibbin and Wilcoxen, 1995) and currently used by the US Environmental Protection Agency (EPA) (McKibbin, *et al.*, 1998).

GEM-E3: a general equilibrium model for the EU-14 (Capros, 1997).

GTEM (Global Trade and Environment Model): dynamic general equilibrium model of the world economy developed by the Australian Bureau of Agricultural and Resource Economics (ABARE: see the web site www.abare.gov.au).

GRAPE: Global Relationship Assessment to Protect the Environment Model, developed by the Institute for Applied Energy, the Research Institute of Innovative Technology for Earth and the University of Tokyo, Japan.

MERGE (a Model for Evaluating the Regional and Global Effects of greenhouse gas reduction policies): an intertemporal market equilibrium combining a bottom-up representation of the energy supply together with a top-down perspective on the remainder of the economy, developed by A. Manne (Stanford University) and R.G. Richels (EPRI). See www.stanford.edu/group/MERGE or Manne *et al.*, 1995.

MS-MRT: a Multi-Sector, Multi-Region general equilibrium Trade model developed by Charles River Associates and T. Rutherford at the University of Colorado.

Oxford Model: a macroeconomic model with a short to medium-term focus.

POLES: a world energy system model used by the European Commission (Criqui *et al.*, 1996).

PRIMES: a partial equilibrium model of the European energy system (Capros and Kokkolakis, 1996).

RICE: Regional Integrated Climate and Economy Model developed by the Yale University.

SGM: Second Generation Model developed by the Batelle Pacific Northwest National Laboratory.

WorldScan: a multi-sector, multi-region, recursively dynamic applied general equilibrium (AGE) model of the world economy, used in a project funded by the Dutch Ministry of Environment (VROM), the Netherlands Bureau of Economic Policy Analysis (CPB), and the National Institute of Public Health and the Environment (RIVM). For a model description, see Geurts *et al.* (1995, 1997).

MOP

Meeting of the Parties (to the Kyoto Protocol): same as COP with non-ratifiers of the Protocol as observers. MOP does not effectively occur until the Kyoto Protocol has been ratified.

QELRC

Quantitative Emission Limitation or Reduction Commitment: permitted emissions for 2008-12 expressed as a percentage of the 1990 level.

Thermohaline ocean circulation

Flows between the equatorial and polar regions, driven largely by relative salinity, carrying heat towards the polar regions. The most important is in the North Atlantic, of warm (near surface) water flowing north and cold (deep) water south. Without this flow the "Gulf Stream" would not have its moderating effect on the climate of western Europe.

UNFCCC

United Nations Framework Convention on Climate Change. Established in 1992 in Rio de Janeiro.

Annex 1

Gain from Trade With Backstop Energy Sources

Under certain circumstances, scenarios implying permit trading, as they are simulated with GREEN, prove to be equally or more costly overall than the corresponding equilibrium with no trade. This is the case in a scenario with emission trading under the Kyoto Protocol in 2050 where, despite 350 million tons of carbon being traded, the total abatement cost of Annex 1 countries with permit trading is slightly higher than with no trading (825 billion 1985 $ instead of 821 billion 1985 $). A similar situation arises in 2050 with the alternative allocation of permits: although trade between the CIS and the other Annex 1 countries would amount to 1 126 million tons of carbon, the equilibrium with trade is globally more costly than the equilibrium with no trade (825 billion 1985 $ against 765 billion 1985 $). These results seem to contradict the widespread view that emission trading reduces total abatement costs.

This note analyses the conditions under which trade reduces total abatement costs. It shows that gains from trade are subject to the condition that the marginal cost curve for the world be concave to the origin (thus its second derivative must be positive). If this condition is not satisfied, a situation involving large amount of trading may prove as or even more costly than the corresponding situation with no trading. In practice, in GREEN, convexities of the world marginal cost curve may occur when backstop technologies become profitable in some countries while other countries still use conventional energy sources.

Consider n countries in a senario with no trading. Their emissions are constrained not to exceed E_i^{NT} for $i = 1...n$ respectively and this is achieved by raising the carbon price to P_i^{NT} in each country i. Alternatively, if countries are allowed to trade emission rights, they set their emissions E_i^T at the level at which their marginal abatement cost is equal to the common equilibrium carbon price P. Trade yields gain if the total abatement cost with emission trading is lower than the total cost with no trade:

$$[1] \qquad P^T \sum_i E_i^T - \sum_i P_i^{NT} \cdot E_i^{NT} < 0 \quad \text{for } i = 1,...n$$

Equation [1] can be rewritten in terms of changes of carbon prices between trade and non-trade cases ($\Delta P_i = P_i^T - P_i^{NT}$) and changes in emission levels between trade and non-trade cases ($\Delta E_i = E_i^T - E_i^{NT}$):

$$[2] \qquad \sum_i \Delta P_i \cdot E_i^T + \sum_i P_i^{NT} \cdot \Delta E_i < 0 \quad \text{for } i = 1,...n$$

Equation [2] can be developed for a two-country case with a country where the marginal abatement cost is high (high-cost country HC) and a country where the marginal abatement cost is low (low-cost country LC). The following table summarises how the cost change due to trade can be decomposed.

Terms	Impact from trade	Sign
[2.1] $\Delta P_{HC} \cdot E^{T}_{HC}$	Marginal cost reduction in the high-cost country.	Negative
[2.2] $P^{NT}_{HC} \cdot \Delta E_{HC}$	Increase of emissions in the high-cost country (which corresponds to the purchase of rights by the high-cost country).	Positive
[2.3] $\Delta P_{LC} \cdot E^{T}_{LC}$	Marginal cost increase in the low-cost country.	Positive
[2.4] $P^{NT}_{LC} \cdot \Delta E_{LC}$	Decrease of emissions in the low cost country (which corresponds to the sale of rights of the low-cost country).	Negative

The expected outcome that trade reduces the overall abatement cost is verified when terms [2.1] and [2.4] are large (and negative). This happens when the slope of the marginal cost curve in the high-country is larger than in the low-cost country. As shown in Figure A1.1, this corresponds to a situation where the world cost curve is concave to the origin. On the contrary, trade may, in principle, yield a larger overall cost if the slope of the marginal cost curve in the high-cost country is smaller than in the low-cost country, as in Figure A1.2 (in this case, terms [2.2] and [2.3] are large and positive). In this case, the cost increase in the low-cost country exceeds the cost reduction in the high-cost country and the right-hand term of equation [2] becomes positive. In GREEN, this occurs when backstop technologies are used in OECD countries while they are still not yet competitive in the CIS (due to the existence of energy subsidies and a different fuel mix). The marginal cost curves reported in Figures A1.3 to A1.8 show that this situation may imply that the cost curve in the OECD area becomes less steeper than in the CIS.

Figure A1.1. **World marginal cost curve is concave to origin**

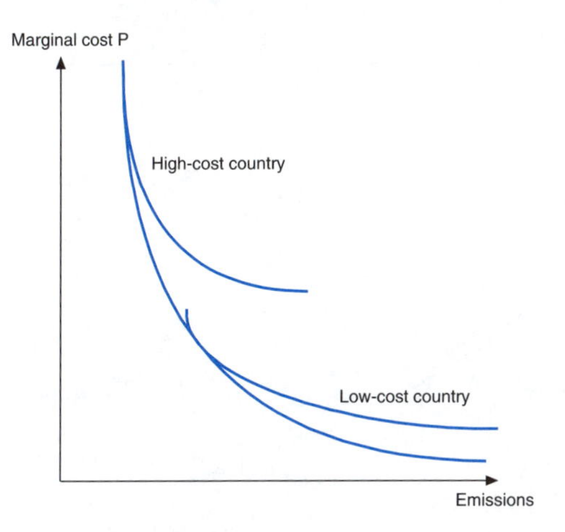

Figure A1.2. **World marginal cost curve is convex to origin**

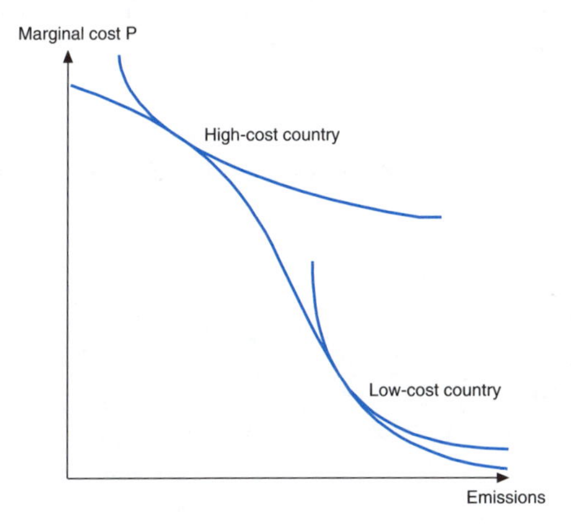

Figure A1.3. **Cost curves for the United States**

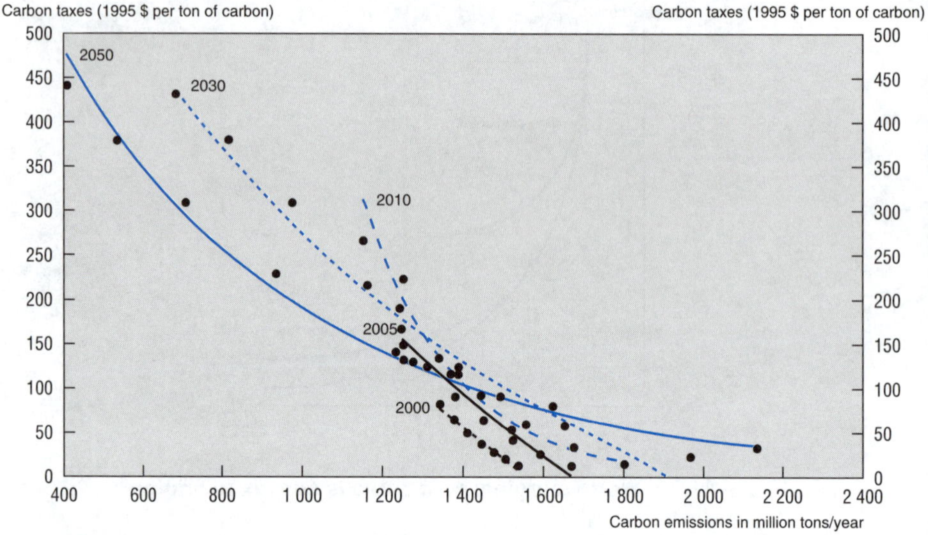

Source: GREEN Model, OECD Secretariat.

Figure A1.4. **Cost curves for the European Union**

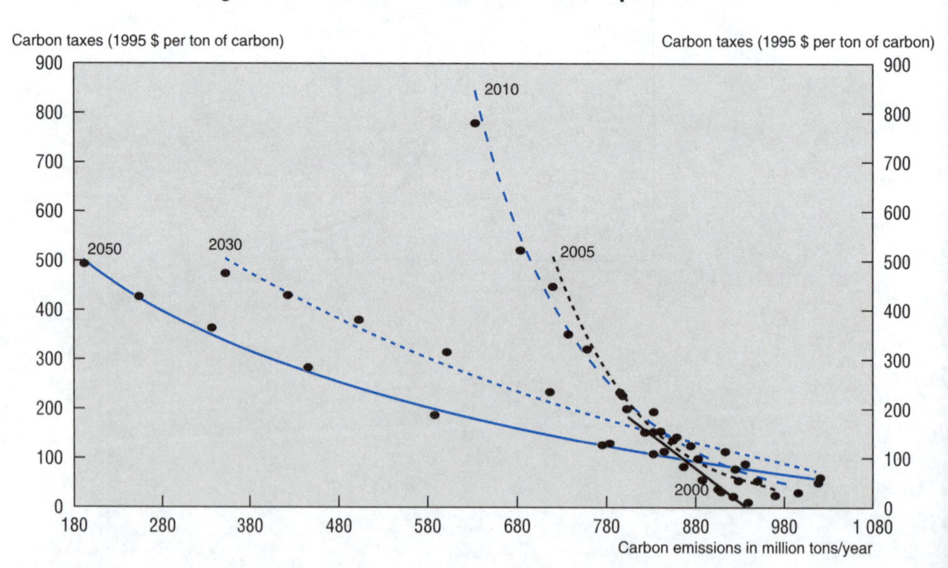

Source: GREEN Model, OECD Secretariat.

Figure A1.5. **Cost curves for Japan**

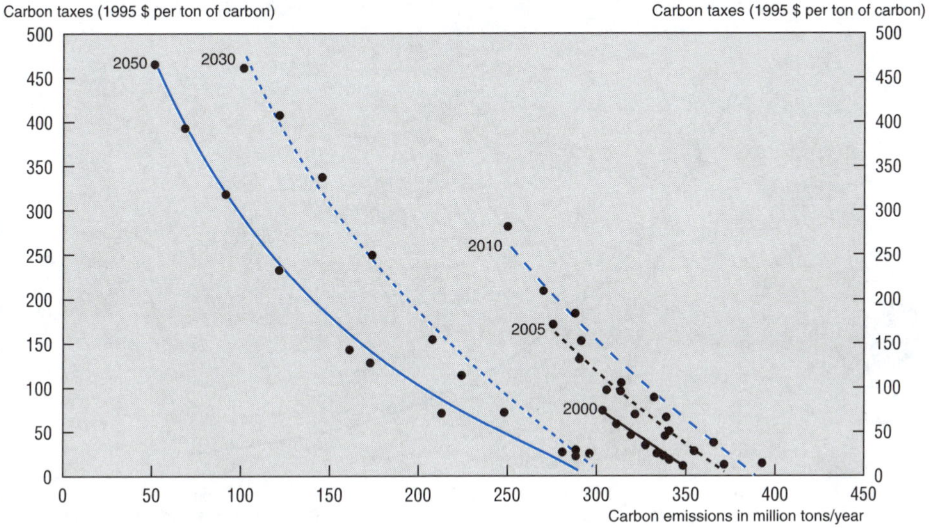

Carbon taxes (1995 $ per ton of carbon)
Carbon taxes (1995 $ per ton of carbon)
500
450
400
350
300
250
200
150
100
50
0
2050
2030
2010
2005
2000
0
50
100
150
200
250
300
350
400
450
Carbon emissions in million tons/year

Source: GREEN Model, OECD Secretariat.

Figure A1.6. **Cost curves for other OECD**

Carbon taxes (1995 $ per ton of carbon)
Carbon taxes (1995 $ per ton of carbon)
500
450
400
350
300
250
200
150
100
50
0
2050
2030
2010
2005
2000
0
100
200
300
400
500
600
Carbon emissions in million tons/year

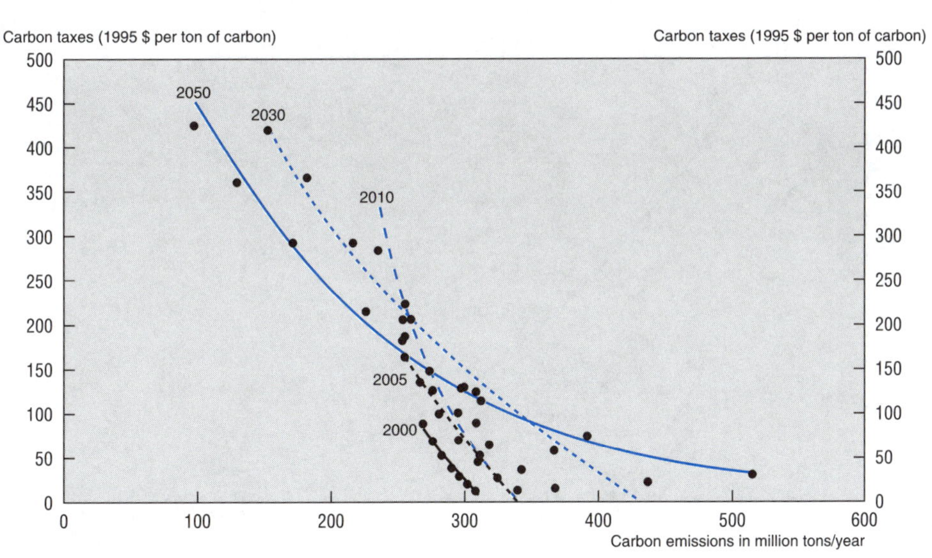

Source: GREEN Model, OECD Secretariat.

Figure A1.7. **Cost curves for Eastern Europe**

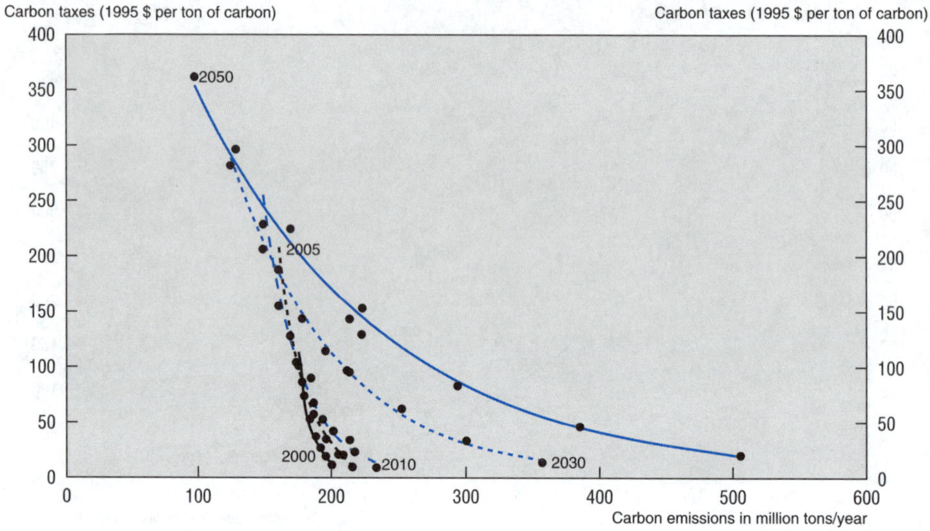

Carbon taxes (1995 $ per ton of carbon)
Carbon taxes (1995 $ per ton of carbon)
400
350
300
250
200
150
100
50
0
2050
2005
2000
2010
2030
0
100
200
300
400
500
600
Carbon emissions in million tons/year

Source: GREEN Model, OECD Secretariat.

Figure A1.8. **Cost curves for the CIS**

Carbon taxes (1995 $ per ton of carbon)
Carbon taxes (1995 $ per ton of carbon)
500
450
400
350
300
250
200
150
100
50
0
2050
2030
2010
2005
2000
50
250
450
650
850
1 050
1 250
1 450
1 650
1 850
2 050
Carbon emissions in million tons/year

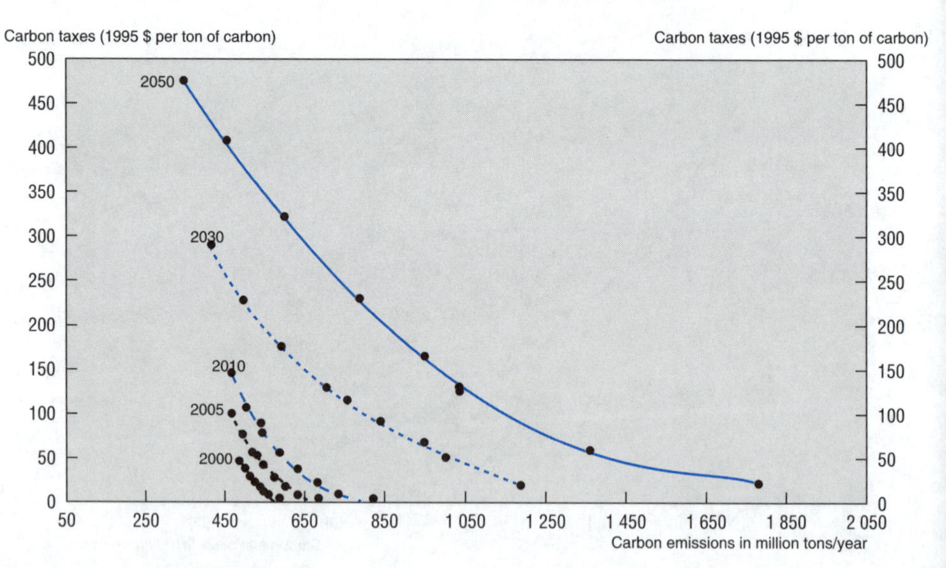

Source: GREEN Model, OECD Secretariat.

Annex 2

Translating Past CO_2 Emissions into Future Concentration

The impact on atmospheric concentrations of policies to reduce emissions is usually assessed by means of reduced-form "response functions". These equations typically show how a change in emissions at a given time would affect the future timepath of concentration relative to the baseline. The baseline itself is obtained from complex models with full specifications of the various components of the carbon cycle. Establishing a baseline for concentration and assessing how policies would modify this baseline is subject to many uncertainties. First, many aspects of the carbon cycle are not well known, including for instance the fertilisation effect, and, more importantly, it is not known how this cycle would behave at concentration levels much above those that were attained in the past.[1] Second, carbon cycles have a very long-term dynamic (several hundred years) which requires projections of economic growth over a period far too long to imagine the technological options that might be developed. In consequence, the following exercise has little predictive value; its aim is only to highlight the dynamics at work.

1. Defining a baseline projection of emissions

This analysis quantifies the impact on concentration of the Kyoto Protocol and alternative policies over a period of 200 years. It covers CO_2 emissions from both fuel combustion and land-use changes. It is based on the emission timepath projected by the GREEN model up to 2050. This timepath is extended for the period from 2050 to 2200 under the following simplified assumptions:

- The population in Annex I countries remains stable after 2050 while, in non-Annex I countries, projections of population are linear extrapolations of the trends over the period 2030-50. As a result, the world population grows by 0.8 per cent annually on average from 2010 to 2050 and by 0.4 per cent hereafter.
- The projection of GDP *per capita* in non-Annex I countries is based on the assumption of a partial catch-up of productivity levels. Following this assumption, the ratio of average GDP *per capita* in non-Annex I countries to Annex I countries is close to 50 per cent in 2200 (against 6 per cent in 1990). At the same time, Annex I GDP *per capita* converges to an average level projected by *linear* extrapolation of the trend over the period 2030-50.
- Emissions *per capita* are projected based on a log-linear relationship between emissions *per capita* and GDP *per capita* which is fitted on data from the BaU scenario from the GREEN model for the period 1995 to 2050. As Figure A2.1 shows, this relationship implies that emissions, as countries grow richer *per capita*, grow less than proportionally

Figure A2.1. **Relationship between emissions per capita and GDP per capita in GREEN BaU scenario**

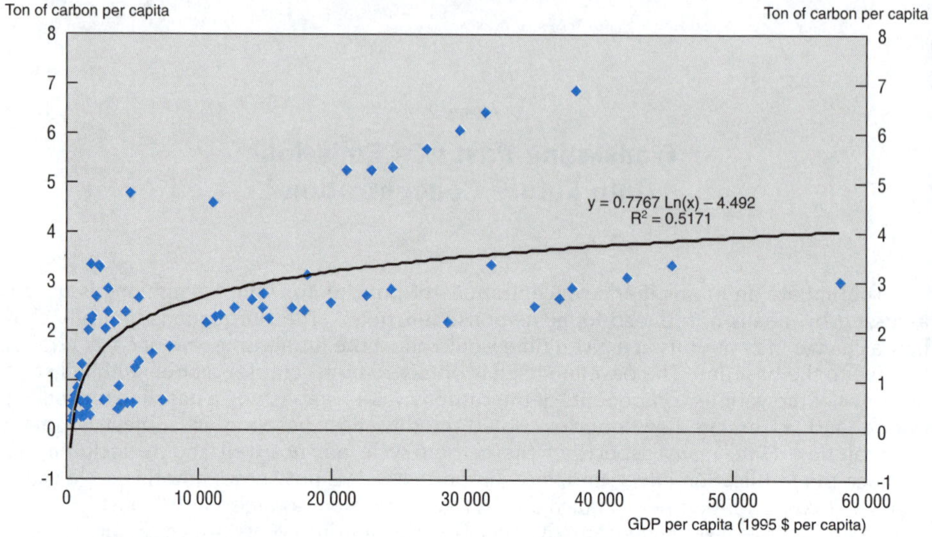

Ton of carbon per capita

$y = 0.7767 \ln(x) - 4.492$
$R^2 = 0.5171$

GDP per capita (1995 $ per capita)

Source: GREEN Model, OECD Secretariat.

with the level of the GDP *per capita*. Emissions *per capita* are assumed to converge to their values predicted by the log-linear relationship by 2200.[2]

Following these assumptions, world emissions are projected to grow 0.7 per cent per year during the period 2050 to 2200 (compared with 2.4 per cent per year from 2010 to 2050, as projected by the GREEN model). Emissions by Annex 1 countries would culminate at 8 gigatonnes per year in 2050 and decline slightly thereafter (by 0.2 per cent per year). Thus, the growth of CO_2 emissions over the very long term would originate entirely from non-Annex 1 countries (mainly the following GREEN regions countries: Energy Exporters, India and Rest of the World). The implication is that, by 2200, the weight of Annex 1 emissions in world emissions would become rather small.

2. Modelling the carbon cycle

2.1. A *reduced-form model of the carbon cycle*

Modelling the carbon cycle is a complex task, as it requires representing a very large number of physical interactions between the oceans, the biosphere and the atmosphere. Therefore, carbon cycle models are usually very large and difficult to use in practice. For the purpose of an economic analysis, it is preferable to use a reduced-form specification – also referred to as an "impulse response function" instead of the original, fully specified, models.

The derivation of the response functions used here is based on a comparison study undertaken as part of the scientific assessment made for the Working Group I of the IPCC (see Enting *et al.*, 1994). The comparison involved 18 models with different levels of complexity, ranging from simple response function models to full general circulation models. An impulse response function is calculated by adding 10 GT of carbon to the background concentration profile and then letting the model calculate the proportion of this additional carbon remaining in the atmosphere after a given period of time. Among the six full models which participated in the IPCC comparison project, three models reported enough information to derive response functions. The concentration calculations presented in this annex are based on these three models. They are:

- The model developed by Wigley (1993) (referred to as Model W) combines response functions of the ocean derived from mechanistic ocean models and parameterised models of the terrestrial systems.
- The model developed by Siegenthaler and Joos (1992) at the University of Bern (referred to as Model J) has fully parameterised terrestrial and ocean components.
- The model developed by Wuebbles and Jain (Jain *et al.*, 1995) at Lawrence Livermore National Laboratory (referred to as Model L) is another fully parameterised model but with a more desegregated treatment of the terrestrial components.

The response function based on all three models are linear and express CO_2 remaining in the atmosphere as a proportion of the amount of CO_2 emitted in previous years. Reduced forms derived from the Wigley model have previously been used in the economic literature about the optimal path of emission reductions (see, for instance, see Ha-Duong, *et al.*, 1997). In the following, the robustness of the results obtained with that model is tested by using the two other models. In addition, response functions are also dependent or the level of the background concentration, as higher concentration reduces the ability of the carbon cycle to sequester carbon.

The specification of the response function is fairly simple (see equation [1]). It expresses the atmospheric concentration in time *t* (C_t) as a function of the weighted sum of past emissions (E_u). The weighting factor – R_{t-u} – characterises the response function; it corresponds to the percentage of CO_2 emitted in time *u* which is still in the atmosphere in time *t*, *i.e.* *t-u* years after it has been emitted (the residual carbon fraction). In principle, its value is equal to unity for the previous year's emissions and declines more or less exponentially for earlier years' emissions. The equation can be written as follows:

[1] $$C_t = C_{t0} + 0.471 \sum_{u=t0}^{t-1} R_{t-u} \cdot E_u$$

with *t*0 being the reference year starting from which concentration are calculated (in practice, 1765 which is considered as the benchmark year for defining the pre-industrial level of concentration). The factor 0.471 converts gigatonnes of CO_2 into ppmv. C_{t0} (the pre-industrial level of concentration) in equation [1] is calibrated so as to reproduce the level of concentration observed in 1990 (354 ppmv).

The response functions derived from the three selected models differ in their values of the residual carbon fraction (R_{t-u}). Figure A2.2 draws the values of these atmospheric carbon fractions over a period of 500 years assuming that the background concentration of carbon in the atmosphere is 650 ppmv.[3] All three functions show a relatively fast rate of decay of carbon during the first 50 years. Thereafter, the proportion of carbon staying in the atmosphere tends to stabilise. The Wigley model has a higher rate of decay over the longer term with the

Figure A2.2. **Three alternative impulse response functions**

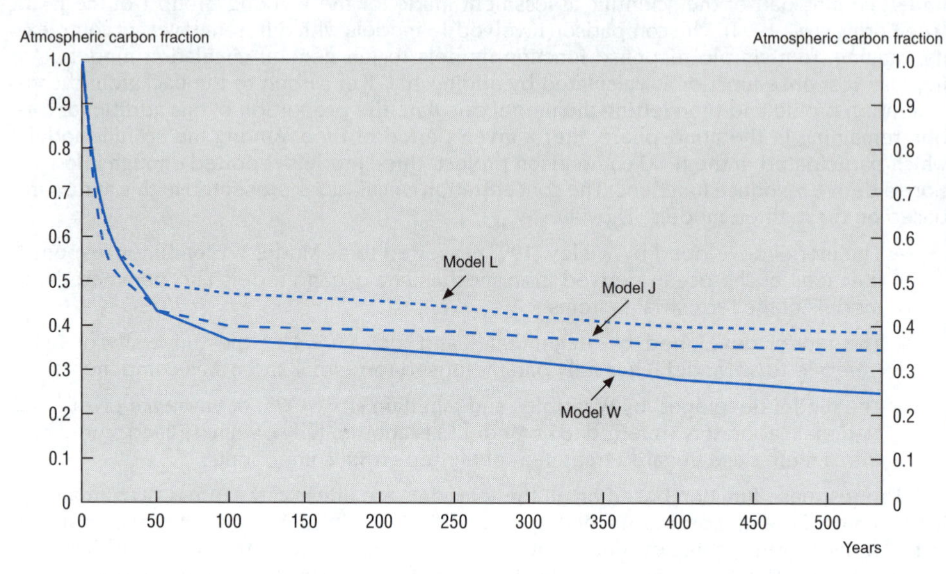

Source: Based on Enting *et al.* (1994).

residual carbon fraction falling to 25 per cent after 500 years. The outcome from the two other models is more pessimistic: even after a period of more than 500 years, around 40 per cent of the carbon emitted now would still be in the atmosphere.

2.2. *Influence of the background concentration level*

The level of the background carbon concentration matters for the shape of the response function, perhaps even more than the choice of model. Simulations made with model L illustrate this. Figure A2.3 shows alternative response functions corresponding to background concentration ranging from 354 ppmv (the 1990 level) to 750 ppmv. The residual carbon fraction in the longer term differs substantially among these functions, ranging from 20 per cent with a background concentration of 354 ppmv to more than 40 per cent with a background concentration of 750 ppmv. Thus, a higher background concentration reduces the ability of the carbon cycle to sequester carbon. The implication is that stabilising concentration at a higher level requires a higher degree of abatement in order to compensate for the lower rate of sequestration.

3. Impact of the Kyoto Protocol

Figure A2.4 shows the evolution of world CO_2 emissions assuming that Annex 1 countries satisfy their commitments for the first budget period (2008-12) of the Kyoto Protocol and then keep their emissions constant at these levels from 2010 to 2200 (the scenario referred to as

Figure A2.3. **Impulse response functions with different background concentrations**

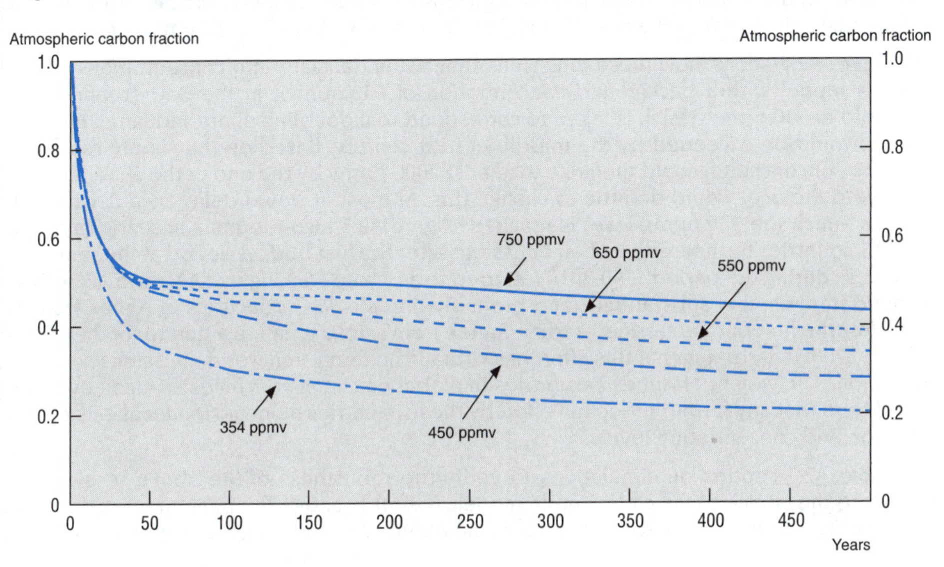

Source: Based on Enting *et al.* (1994).

Figure A2.4. **Long-term impact of the Kyoto Protocol on CO_2 emissions as projected by the GREEN model**

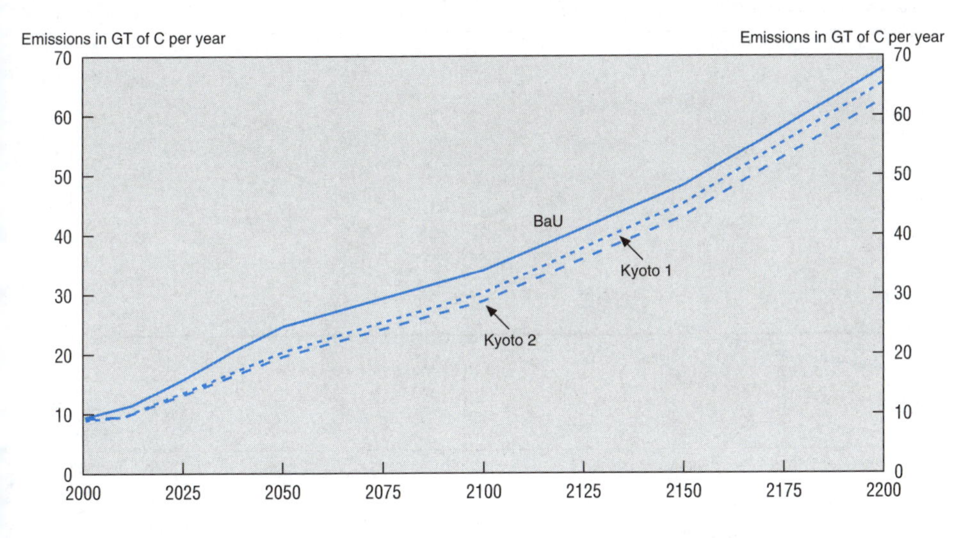

Source: GREEN Model, OECD Secretariat.

Kyoto 1 in Figure A2.4). This timepath is compared with the BaU scenario, as defined above. The impact of the Protocol on world emissions thus appears very moderate and even declining after 2100, in line with the diminishing importance of Annex 1 in total emissions.

Figure A2.5 shows how these emissions timepaths translate into concentrations (using Wigley's model with a background concentration of 650 ppmv). In the BaU scenario, the threshold of 550 ppmv, which is taken to correspond to a doubling of pre-industrial concentration, would be exceeded by the middle of next century. Based on the simple response function, concentration could then rise to almost 2000 ppmv by the end of the 22nd century. The Kyoto Protocol would do little to change this. At most, it would delay by a decade the point at which the 550 ppmv level is reached. Figure A2.5 also reports a scenario in which Annex 1 countries further reduce their emissions after the first budget period at the same linear rate as during the period 1990-2010 (referred to as Kyoto 2 on Figures A2.4 and A2.5). This scenario implies very substantial – and probably unrealistic – reductions in Annex 1 countries after the first budget period, leading Annex 1 emissions to fall to a quarter of their 1990 level in 2200. It is striking that this effort has virtually no impact on world emissions and concentration. The weight of Annex 1 countries in global emissions becomes so small over the longer term that any further reduction effort by these countries alone no longer affects global emission and concentration levels.

Table A2.1 reports on simulations to verify the robustness of the above results with regards to the choice of the carbon cycle model. The table shows the percentage reductions of CO_2 concentrations in the two scenarios of the Kyoto Protocol using the response functions

Figure A2.5. **Long-term impact of the Kyoto Protocol on CO_2 concentration[1] as projected by the GREEN model**

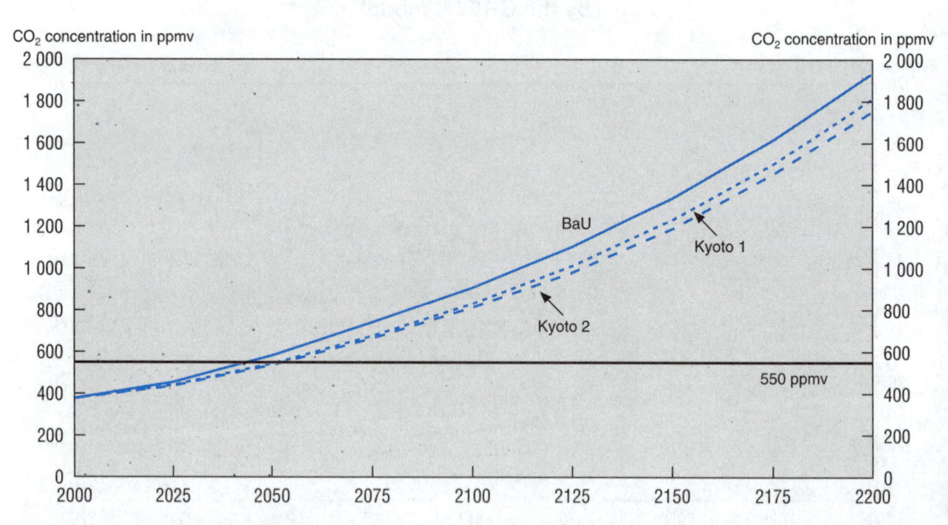

1. Using Wigley's model (Model W).
Source: GREEN Model, OECD Secretariat.

Table A2.1. **Concentration reduction resulting from the Kyoto Protocol: Comparison across carbon cycle models**

Per cent deviation relative to BaU

	Kyoto 1		Kyoto 2	
	2100	2200	2100	2200
Model W	–11.0	–7.9	–13.3	–11.0
Model J	–10.7	–8.1	–13.0	–11.2
Model L	–11.1	–8.4	–13.4	–11.5

Kyoto 1: Annex 1 countries satisfy their commitments for the first budget period (2008-12) of the Kyoto Protocol and then keep their emissions constant at these levels to 2200.
Kyoto 2: Annex 1 countries keep their emissions declining after the first budget period at the same linear rate as during the period 1990-2010.
Source: OECD Secretariat, based on Enting *et al.*, 1994.

derived from the three carbon cycle models described above. All three models project the resulting reductions of CO_2 concentration to be around 10 per cent.

4. Stabilisation scenarios

Figures A2.6 and A2.7 show the results from a scenario in which, after the first budget period of the Protocol, Annex 1 and non-Annex 1 countries keep their CO_2 emissions constant at their 2010 levels (the scenario called NA1 stabilisation). The impact on concentration is substantial (as calculated by the Wigley model with a background concentration of 650 ppmv): it would delay by half a century the time at which the 550 ppmv limit would be exceeded. But, as Figure A2.7 shows, concentration would still be increasing by this time and would reach 720 ppmv in 2200. Furthermore, the scenario is not likely to be politically viable, given that, by 2010, non-Annex 1 emissions per capita would still lie well below that in Annex 1 countries (0.7 ton of CO_2 per year and per capita in non-Annex 1 countries on average against 3.6 tons per capita in Annex 1 countries). Past and current negotiations about climate change suggest that any agreement which does not satisfy to some kind of egalitarian principle has little chance of being accepted by a large number of developing countries.

One particular example of an egalitarian distribution of emission rights is that non-Annex 1 countries impose emissions limitation only when their emissions *per capita* ratio reach those of Annex 1 countries. The implication of this assumption is that the pace at which Annex 1 countries reduce their own emissions will determine the willingness of non-Annex 1 countries to limit their emissions. More rapid and stringent reductions of emissions in Annex 1 countries reduce the ability of non-Annex 1 countries to increase their emissions.

The impact of the Kyoto Protocol based on this distribution key, *i.e.* by assuming that non-Annex 1 countries do not exceed the *per capita* emissions in Annex 1 countries and that these countries keep their emissions constant at the levels specified in the Protocol, is shown in Figures A2.6 and A2.7 (scenario Egal 1). In this case, the Kyoto Protocol contributes much more to reduce emissions and concentration than if non-Annex 1 countries are free from any obligation. However, the scenario does not achieve any kind of stabilisation. Assuming conditional action by non-Annex 1 countries in the case where Annex 1 countries set their emissions to decline after the first budget period makes an even larger difference:

Figure A2.6. **Alternative strategies of emission reductions involving non-Annex 1 countries**

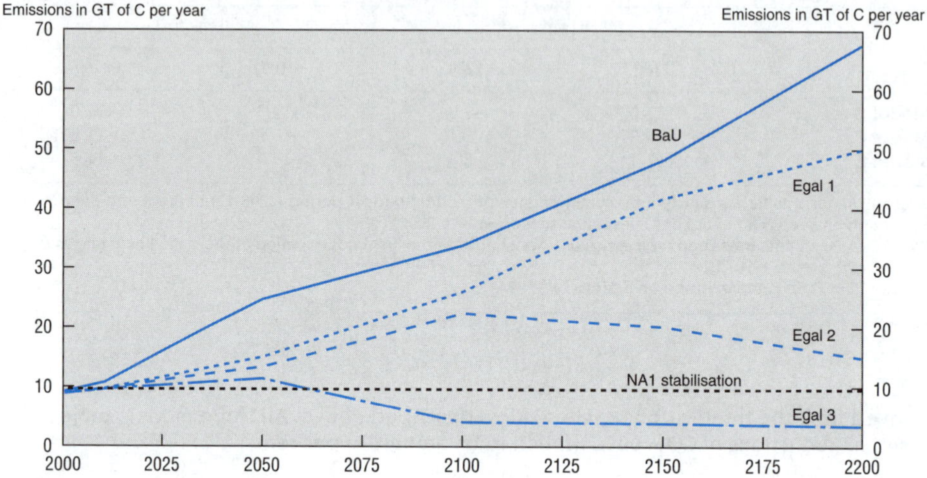

Emissions in GT of C per year
Emissions in GT of C per year
70
60
50
40
30
20
10
0
2000
2025
2050
2075
2100
2125
2150
2175
2200
BaU
Egal 1
Egal 2
NA1 stabilisation
Egal 3

Source: GREEN Model, OECD Secretariat.

Figure A2.7. **Long-term impact of emission reductions in non-Annex 1 countries on concentration**[1]

CO_2 concentration in ppmv
CO_2 concentration in ppmv
2 000
1 800
1 600
1 400
1 200
1 000
800
600
400
200
0
2000
2025
2050
2075
2100
2125
2150
2175
2200
BaU
Egal 1
Egal 2
NA1 stabilisation
Egal 3
550 ppmv

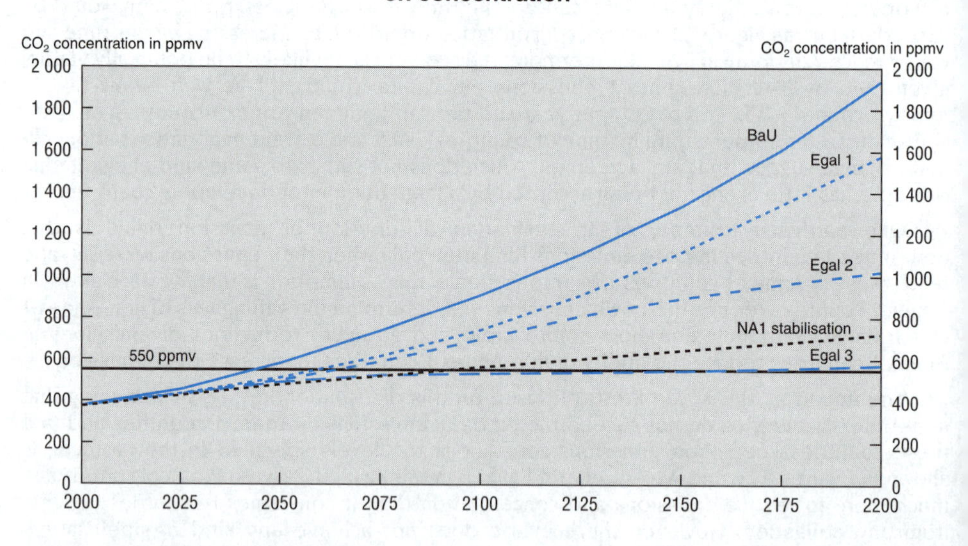

1. Using Wigley's model (Model W).
Source: GREEN Model, OECD Secretariat.

it would cause non-Annex 1 countries to start reducing their own emissions by 2100, with concentration nearing stabilisation at around 1000 ppmv in 2200 (scenario Egal 2 on Figures A2.6 and A2.7).

Figure A2.6 also shows a scenario in which CO_2 concentration is stabilised at 550 ppmv (Egal 3). In this scenario, Annex 1 countries set a target for their own emissions in 2100 equal to 40 per cent of the average *per capita* emissions in the non-Annex 1 countries in 1990. This target implies very drastic cuts: by 2050, Annex 1 emissions are bound not to exceed half of their 1990 level. Non-Annex 1 countries are assumed to impose restrictions as soon as their emissions *per capita* exceed the average for Annex 1 countries. As a result, they undertake a serious mitigation effort starting from 2010 and, by 2050, their emissions need to be cut by half relative to their BaU level.

This last scenario confirms that any strategy to stabilise the CO_2 concentration at below 550 ppmv would require that world emissions decline to levels which are well below current emissions (in Egal 3, world emissions fall to 4 gigatonnes in 2100 compared with 8 gigatonnes in 1990). This, of course, cannot be achieved without the participation of non-Annex 1 countries.

5. Steady-state stabilisation versus local maxima

Any single stabilisation target can be reached by a variety of alternative emission timepaths. As discussed in Chapter 3, the choice is between taking early measures in order to spread the burden of the adjustment over a longer period of time or, alternatively, delaying the adjustment in the hope that it can be made at lower economic costs later on. Figures A2.8*a* to 8*d* aim at highlighting the long-term properties of the carbon cycle dynamic. They show different concentration timepaths (based on Model W with a background concentration of 650 ppmv) corresponding to reductions to given minimum levels of emissions taking place over different time spans. For instance, in Figure A2.8*a*, all trajectories refer to emission reductions spread over the period from 2010 to 2030, with emissions remaining constant thereafter. Specifically, the trajectory 4 GT corresponds to a reduction of world emissions down to 4 gigatonnes of carbon per year (compared with 8 gigatonnes in 1995) while the timepath 0 GT simulates the complete elimination of any CO_2 emissions over the same period. A number of findings come from these diagrams:

- Only a few trajectories lead to effective stabilisation of CO_2 concentration over the long run. These trajectories (in bold in Figures A2.8*a* to 8*d*) correspond to an equilibrium in which emissions match the amount of carbon which is sequestered from the atmosphere in each year. The level of these equilibrium emissions is very low: it amounts to 0.5 gigatonnes of carbon per year if the stabilisation starts in 2010 (Figures A2.8*a* and 8*c*) and to 1 gigatonne of carbon per year roughly if the stabilisation starts in 2050 (Figures A2.8*b* and 8*d*). Thus, effectively stabilising the CO_2 concentration requires bringing emissions down to very low levels compared with the current ones.
- Trajectories with long-term minimum emissions above these equilibrium levels do not effectively lead to stable concentration in the sense that, in the long run, concentrations are still increasing. However, some of these trajectories exhibit a local maximum after which concentrations decline temporarily and then start rising again (for instance, 4 GT in Figure A2.8*b*). Emission reductions corresponding to these trajectories are, at the most, seen as delaying the rise of concentration rather than stabilising them.

Figure A2.8. **Alternative pathways to stabilise CO_2 concentration**[1]

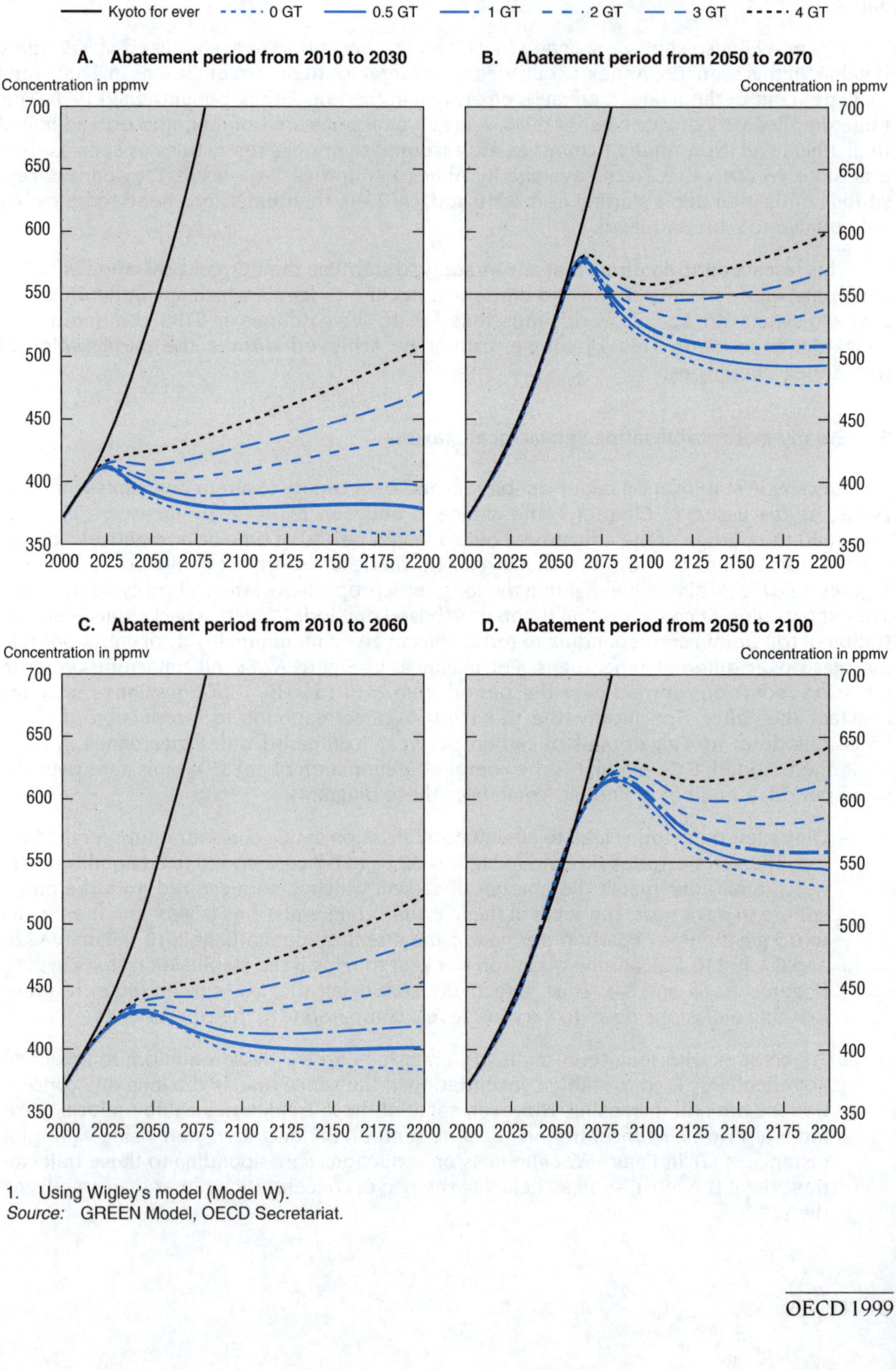

Kyoto for ever | 0 GT | 0.5 GT | 1 GT | 2 GT | 3 GT | 4 GT

A. Abatement period from 2010 to 2030

B. Abatement period from 2050 to 2070

Concentration in ppmv

700, 650, 600, 550, 500, 450, 400, 350

2000 2025 2050 2075 2100 2125 2150 2175 2200

C. Abatement period from 2010 to 2060

D. Abatement period from 2050 to 2100

Concentration in ppmv

700, 650, 600, 550, 500, 450, 400, 350

2000 2025 2050 2075 2100 2125 2150 2175 2200

1. Using Wigley's model (Model W).

Source: GREEN Model, OECD Secretariat.

- Delaying reductions increases the equilibrium level of concentration (*i.e.* the level at which the concentration does not rise any more). Concentration could be stabilised at 380 ppmv if reductions start in 2010 (0.5 GT in Figure A2.8*a*) while the corresponding level is 513 ppmv if there is no reduction before 2050 (1 GT in Figure A2.8*b*).
- Spreading the reductions over a large time-span increases the equilibrium level of concentration as well as their local maximum, increasing the risk of irreversible climate damage (compare, for instance, 1 GT in Figure A2.8*d* with 1 GT in Figure A2.8*c*).
- Non-Annex 1 countries may effectively wait until 2050 before imposing any reduction (as in Figures A2.8*b* and A2.8*d*) as long as world emissions can be cut down to less than 1 gigatonne per year over a more or less long period of time. A similar conclusion is reported by Wigley *et al.* (1996).
- A strategy of early but relatively small cuts that leave emissions above their long-term equilibrium levels may do less for stabilising the climate than a strategy of late but very drastic cuts. This can be seen by comparing the trajectory stabilising emissions at 4 GT over the period 2010-2030 (Figure A2.8*a*) with the trajectory stabilising emissions at 1 GT over the period 2050-2070. However, the former would maintain concentration at lower levels on average over the period considered although still on a rising trend. The former strategy is likely to be more costly, but the latter implies a higher degree of risk as concentration approaches the 550 ppmv threshold much more rapidly.

The prerequisite for obtaining steady-state stabilisation is that annual emissions not exceed the natural rate of carbon sequestration. Table A2.2 reports the level of emissions corresponding to a steady-state stabilisation with the three alternative models assuming a background concentration of 650 ppmv. All three models yield very low (below 1 GT of carbon per year) and rather comparable equilibrium levels of emissions. Equilibrium emissions from models J and L are lower than with model W, as one would expect from the higher residual carbon fraction over the longer term in these two models (Figure A2.2).

Table A2.2 also reports equilibrium emissions with a background concentration of 450 ppmv (using model L); their levels are higher than with the corresponding 650 ppmv response functions, as the residual fraction of carbon is smaller when the background con-

Table A2.2. **Equilibrium levels of emissions corresponding to concentration stabilisation: Comparison across carbon-cycle models**

Gigatonnes of carbon per year

	Time frame of emission reductions			
	2010-2030	2010-2060	2050-2070	2050-2100
Model J (650 ppmv)[1]	0.2 (383)[2]	0.3 (409)	0.4 (518)	0.4 (575)
Model L (650 ppmv)	0.3 (397)	0.4 (431)	0.6 (559)	0.7 (628)
Model W (650 ppmv)	0.4 (372)	0.5 (399)	0.8 (500)	1.0 (555)
Model L (450 ppmv)	0.7 (386)	0.9 (417)	1.8 (544)	2.3 (612)

1. Figures in parenthesis refer to background concentration levels.
2. Figures in parenthesis report the corresponding levels at which concentration is stabilised.

centration is lower (see Figure A2.3). Thus, with early action (from 2010 onward), the background concentration remains lower and emissions may be stabilised at a higher equilibrium level (around 1 GT per year). Later action, by letting concentration rise more, would require emission cuts to a somewhat lower level (0.6-0.7 GT per year).

Notes

1. A recent study by Cao and Wooodward (1998) suggests that the fertilisation effect would level off at a CO2 concentration of 450 ppmv. If true, this would imply almost no fertilisation effect for the range of concentrations which are considered in the following scenarios..
2. For some Annex 1 countries (for instance the United States and the CIS), this implies a long-term decline of the emissions *per capita* ratio.
3. As carbon concentration is projected to rise from its current level of 370 ppmv to around 900 ppmv by the end of the next century, it is justified to use an average value of 650 ppmv for the background concentration.

Annex 3

Defining Alternative Burden-Sharing Rules

This annex displays the specification of the three alternative burden-sharing rules.

1. "Ability to pay" rule

Annex1 emissions in time t :

if $r \in Annex1 \qquad \overline{E}_{r,t} = (1-\gamma) \cdot E_{r,t0} < E_{r,t}$

with γ *being a given rate of reduction;* $E_{r,t}$ *and* $E_{r,t0}$*, the unconstrained emissions in times* t *and* $t0$*; and* $\overline{E}_{r,t}$*, the constrained emissions in time* t.

Non – Annex 1 emissions in time t :

$$if \quad r \in Non-Annex1 \quad and \quad \left(\frac{GDP_{r,t}}{POP_{r,t}}\right) \geq \alpha \left(\frac{GDP_{Annex1,t}}{POP_{Annex1,t}}\right)$$

$$Then \quad \overline{E}_{r,t} = E_{r,t0} \cdot \left(1 + \gamma_{r,t}^{BaU} - \varepsilon\left(\frac{GDP_{r,t} / POP_{r,t}}{GDP_{NAnnex1,t} / POP_{NAnnex1,t}}\right)\right)$$

with $GDP_{r,t}$, $GDP_{Annex1,t}$ *and* $GDP_{NAnnex1,t}$, *being the GDP of the non – Annex1 country r, of total Annex1 countries and total non – Annex1 countries;* $POP_{r,t}$, $POP_{Annex1,t}$ *and* $POP_{NAnnex1,t}$ *being the populations of the non – Annex1 country r, of the total Annex1 countries and total non - Annex 1 countries;* $\gamma_{r,t}^{BaU}$ *being the emissions growth rate in the Business – as – Usual scenario; and,* α *and* ε*, being calibrated parameters.*

2. "Equal per capita emissions" rule

Annex1 emissions in time t :

if $r \in Annex1$ $\quad \overline{E}_{r,t} = (1-\gamma) \cdot E_{r,t0} < E_{r,t}$

with γ *being a given rate of reduction;* $E_{r,t}$ *and* $E_{r,t0}$*, the unconstrained emissions in times t and t0; and* $\overline{E}_{r,t}$*, the constrained emissions in time t.*

Non – Annex 1 emissions in time t :

if $r \in Non-Annex1$ $\quad \overline{E}_{r,t}$ $\quad$ *such as* $\quad \dfrac{\overline{E}_{r,t}}{POP_{r,t}} \leq \dfrac{\overline{E}_{Annex1,t}}{POP_{Annex1,t}}$

and $\quad \overline{E}_{r,t} < E_{r,t}$

with $\overline{E}_{r,t}$ *and* $\overline{E}_{Annex1,t}$*, being the emissions in the non – Annex1 country r and Annex1 countries;* $POP_{r,t}$ *and* $POP_{Annex1,t}$ *being the populations in the non – Annex1 country r and the Annex1 countries.*

3. "Grandfathering" rule

World emissions in time t : $\overline{E_t} = (1-\gamma) \cdot E_{t0} \leq E_t$

with γ *being the rate of reduction relative to the world emissions in* $t = 0$;

E_t *and* E_{t0}*, the unconstrained world emissions in time t and t0;*

$\overline{E_t}$*, the constrained world emissions in time t.*

Country r emissions in time t : $\overline{E}_{r,t} = share_{r,t0} \cdot \overline{E_t}$ $\quad$ *for* $r = 1, N$ *country / regions*

with $share_{r,t0}$ *being the share of country r in world emissions in* $t = 0$;

$\overline{E}_{r,t}$*, the constrained emissions of country / region r in time t.*

OECD PUBLICATIONS, 2, rue André-Pascal, 75775 PARIS CEDEX 16
PRINTED IN FRANCE
(11 1999 03 1 P) ISBN 92-64-17113-4 – No. 50839 1999